IN OR OUT OF THE MILITARY

To Russ —
Thanks for always taking
this project seriously when I
couldn't quite.
With warmest regard,
Diane

☆ ☆ ☆ ☆

IN OR OUT OF THE MILITARY

HOW TO MAKE YOUR OWN BEST DECISION

D. F. REARDON, PH. D.

•PEPPER PRESS
Oak Harbor, Washington 98277-3288

Cover designed by Roxanne Slimak, New York.
Layout by Baypointe Press, Corona del Mar, California.
Printed by Emerald City Graphics, Incorporated, Kent, Washington.
This text of this document has been printed on recycled paper.

Published by:

• P E P P E R   P R E S S

1254 West Pioneer Way, Suite A266
Oak Harbor, WA 98277-3288

The information in this book reflects the author's experiences and research and is not intended to replace professional vocational or psychological counseling, or medical or legal advice. The purpose is to organize and make available decision-making tools and to provide access to other sources of information. The author and Pepper Press shall have neither liability nor responsibility to any person with respect to loss or damage alleged to be caused directly or indirectly by the information presented. Descriptions and examples of personal decision making are composites to protect individual privacy. Reactions, corrections, and suggestions for future editions are welcome by the author, c/o Pepper Press.

Copyright © 1993 by D. F. Reardon, Ph.D.
First Printing 1993.
Printed in the United States of America.

Quotes from Ira Progoff with permission of the author.
Quotes, adaptations of flow charts, and Trial Balance Sheet from Irving Janis and Leon Mann by permission of the author Irving Janis.
Drawing "The Rest of Your Life" (Figure 6–1) adapted by permission of the artist from "Don't Look Back" 100 ME-0100, Mary Engellbreit Co., 1992.
From *The Tao of Pooh* by Benjamin Hoff. Copyright ©1982 by Benjamin Hoff; text and illus. from *Winnie-the-Pooh* and *The House at Pooh Corner*, CR 1926, 1928 by E.P. Dutton, ©1953, 1956 by A.A. Milne. Used by permission of the publisher, Dutton, an imprint of New American Library, a division of Penguin Books USA Inc.
Excerpts from "The High Desert" from *The Right Stuff* by Tom Wolfe. Copyright ©1979 by Tom Wolfe. Reprinted by permission of Farrar, Straus & Giroux, Inc.
Lifemap ©1993 by D.F. Reardon, Ph.D.

Reardon, D. F.
    IN OR OUT OF THE MILITARY
    How to Make Your Own Best Decision
    Illustrated.
    Includes index.
    Library of Congress Catalog Card Number: 92-061683
    ISBN 1-882287-44-4

This book is dedicated to my father, Albert McGunigle, who supported my decisions, however I made them.

ACKNOWLEDGMENTS

I want to thank my family, friends, and acquaintances for their patience during my long involvement with the research and writing of this book. Special thanks to Barbara, Barbara, and Mary. Thanks also to the many civilians and active duty and retired military in the NAS Whidbey/Oak Harbor community for their helpfulness, good humor, and energetic follow-through. In particular, thanks to Kathy Hunter, Vickie and Darrell Jones, Jim Viskoshill, Linda Harrington, Marie Richmond, Sue Porritt, Boyd Woolsey, and, in memory, Lisa Heva.

In addition to the many workers in the field of decision making, I am grateful for help at critical points from Eula Palmer, James Friedrich, Sean Fagan, B. G. Sick, Vicki L. Sears, Robert Olds, Lee and Hugh Brainard, Sara and Matt Clark, and Roxanne Slimak. Others helped keep life manageable as writing went on. Ken Sebens, Paige McGuire, Diane Rottler, and Patti Carter with the staff of Pony improved my outer world. Nancy Minter, Jim Patterson, David Roomy, Stan Williams, and Grady Grady were companions in the inner worlds of book making.

For support throughout, I thank my husband, Paul, whose actions supported this work of words, and, for inspiration and practical help from beginning to end, I thank my brother, Richard McGunigle.

☆　　☆　　☆　　☆

TABLE OF CONTENTS

INTRODUCTION

Deciding to stay in or get out of the military is one of the biggest decisions of your life.

It's a big decision because it can lead to changes, to many endings, many beginnings, and all the thoughts and feelings that go with them. This book is about the how and why of deciding, how to make your own best decision, and why it's so easy to get stuck partway through with no decision at all.

This book is directed to the military person who is considering leaving, but also to family and friends who care about the decision because it will affect them as well. For example, if you're a spouse whose mate keeps talking about "getting out" without doing anything about it, the indecision can play havoc with plans you try to make. It's hard to commit to activities that last even a year when you're living in such an unsettled situation.

As a psychologist in a rural town with a naval air station, I have drawn on the experiences of Navy personnel as they've struggled with the decision of whether to stay or go. I have also drawn on research studies on decision making in three areas: game theory, risk analysis, and belief systems. Game theory looks at how people balance costs and benefits in decisions like gambling and buying insurance. Risk analysis studies how people compare losses and gains in a one-time choice like buying a car. Studies of belief systems focus on religious and political issues.

Although I use these types of research, I recognize their limits. They don't cover the long-term commitments involved in personal life decisions. Life and career decisions are more complex than any one set of research ideas, so I've selected and adapted the ideas and tools from the three kinds of research. You will learn how to select and adapt them to your situation in a decision-making approach that suits you.

For some of you, the main purpose of reading this book is to help justify a decision you have already made. If you already know what you want to do, these ideas may help you rationalize the decision to yourself. You may also find them helpful in explaining your decision to others.

Some of you aren't sure about what to do or when to do it, but have always trusted your intuition to tell you what to do when the time is right. Many of our biggest life decisions are made this way. They appear to be made when we're not looking. When these decisions surface, we are as surprised as anyone. If you suspect that you reach decisions in this way, more underground than above, you may want to use the techniques in this book to help create a sense of order and control as you wait for your decision to form. Using the tools in this book will occupy your mind and hands while you wait, and the information you gather will provide a useful action map once you know what you want to do.

> *Patience is bitter*
> *but its fruit is sweet*
> *— John Jacques Rousseau*

There are many books on mid-life crises and job stress and career change. Those I think are particularly helpful are listed in Appendix A. A more comprehensive listing is available from career planning seminars and local colleges. Many give super help with what to do once you've decided to leave a career, but few have highlighted how to decide when to leave a job. *None has addressed the needs of those contemplating leaving the military.*

In addition to the large number of books available on career changes, there are many about decision making in general, but most of them have disadvantages for military personnel. Some mix up little decisions and big decisions, treating whether to buy a carpet just like deciding when to leave the military. Others are good for a younger military person but not someone approaching 20-year service, or vice versa. I recognize that there is a strong pull to go for 20 years once you're past 12 to 15 years, but the approach presented here will help no matter how long you've been in or when you are considering getting out.

Research on decision making is taking place in management, medicine, law, and international relations as well as in psychology. I have tried to sample the approaches from these fields, but my selections probably reflect my biases. For example, although I've included the most recent material available, I suspect that I've done this more competently in the psychology areas than in others, since that's my field. Some of you know more about mathematical decision models used in computer simulation studies than I ever will. Add your own knowledge, biases, and notes in the margins, and please drop me a line telling me of areas I have missed; I will include them in future revisions.

The aim of this book is to give you:

    1) a clear set of decision steps

    2) for making major life decisions,

    3) based on updated research information.

I hope this book helps you and the people close to you through the struggle of deciding when to stay and when to go. Whether it's your decision, one shared with others, or one you want to support, understanding the expected and unexpected pitfalls of the process may ease the way.

0.1 – HOW TO USE THIS BOOK

I've arranged the book to follow the steps that usually occur when real life decisions evolve over time.

Step 1   shows you how to take stock of your current situation by analyzing how change and decision making have played a role in your life (Chapter 1).

Step 2   lays out your Options in ways that help, rather than hinder, decision making (Chapter 2).

Step 3   walks you through the anticipated PROs and CONs of each Option in a straightforward way (Chapter 3).

Step 4   sums up your information and handles the close judgment calls you'll have to make (Chapter 4).

There are times when circumstances or your own decision-making styles get in the way of going smoothly through the four steps. Turn to Chapter 5 – Decision Traps and Escape Hatches for help. If you've already been feeling stuck about your decision, understanding such traps is crucial.

The final chapter (Chapter 6) helps you move from making your decision into the new territory of implementing it.

Just as there's no one way to make major life changes, there's no one way of deciding. My clients have taught me that every time I think I've got a single system for helping everyone with life decisions, there's something that doesn't fit for a particular person. They have taught me that each person is truly unique, and has a unique pattern of making decisions as well. You may want to read through the book quickly, stopping and marking the parts that really hit home. You may find that reading the chapters in order works well for you or you may decide to read the chapters in your own order.

0.2 – EXERCISES

There are three major written exercises in the whole book and they are the first three of your four steps:

• Lifemap in Chapter 1,

• Table of Options in Chapter 2, and

• Trial Balance Sheet in Chapter 3.

All places in the text where you are asked to stop reading and work on the exercises are set off with stars, like this:

☆ ☆ EXERCISE ☆ ☆

These exercises certainly will be easier if you do them in pencil rather than pen. Your first tentative guesses serve the same purpose as the first entries in a crossword puzzle: pencil your ideas in lightly to begin building the overall interlocking pattern. Keeping things erasable helps you use your first shaky estimates and guesses until your pattern of deciding begins to take clearer shape. For example, you'll probably want to go back to add or change entries to the Lifemap exercise in Chapter 1 after reading the ideas in later chapters.

If you absolutely detest writing things down, if you shudder at the idea of making lists, don't do it. There are reasons you avoid such things and this is an important part of your personal decision-making style. Read through the exercises anyway, thinking through your responses. It may help to summarize your thoughts in your own way for the Lifemap, Table of Options, and the Balance Sheets. If you don't like writing at all, continue reading and thinking through the exercises.

If you are frustrated by the limitations of the exercise sheets, you may enjoy getting your thoughts down on paper. If you like to write your thoughts out, get yourself a notebook so you can keep your information all in one place. If you find you run into a temporary writer's block, you may find the "hot pencil" technique useful.*

When you see an asterisk (*) in the text, this means that details and examples that expand on the subject are included in a box at the bottom of the page. The box at the bottom of this page describes the Hot Pencil Technique, and is keyed to the subject in the preceding paragraph by the asterisk at the end of the last sentence. These boxes contain material "For Your Information" and are not critical to the sequence of the chapters.

---

* The Hot Pencil Technique: You may like writing but worry about finding yourself with nothing to write. All of us at times find our thoughts blocked and our pencils immobilized by the presence of a sheet of white paper. Whenever this happens, you can use the "hot pencil" technique. If you freeze up, keep the pencil moving off in the margin and keep it writing, writing anything at all, as long as you keep that pencil moving. Don't let it cool down. What you write may be something like: "I can't think of anything to write, there's nothing coming into my head, I can't do this, this is stupid...etc., etc." Just keep the pencil moving until you realize you wrote an item that is useful, then shift back and enter it on the main page.
"I don't know if I'll finish this exercise before dinner. Maybe I can catch that rerun of..."
"Oh, yeah, I'll put in how I want a new VCR when I retire..."

CHAPTER 1

STEP 1 — TAKING STOCK

Let's face it. Deciding to make changes is a hassle.

There's a guaranteed amount of stress and upheaval with change, and military families know this all too well. Some changes end up in unhappiness and some in delight, but any change requires work. It's just human nature to stall before any change can be made, even one we're delighted about.

Making a change is hassle enough when you're sure of what you want. When you're not sure, it's even harder. In deciding to stay in or leave the military, for example, you have to go over in your mind all the PROs and CONs. The longer you've been in, the more complex and confusing this can be. As you try to think about it carefully, you can get sidetracked by one aspect of the problem and lose sight of the whole picture. Trying to keep all these aspects in mind gets complicated, and can become overwhelming very quickly. The natural reaction is to give up trying to think through your decision in a rational way and even avoid thinking about it altogether.

You wouldn't be the first person to turn in your papers because you're tired of trying to think it through, because you just want to get on with your life. You wouldn't be the first person to decide to get out because you couldn't stand the indecision. The only problem with deciding "just-to-decide" is that when negative consequences come along later, which is inevitable for any complex decision, they are harder to handle because you haven't prepared for them. What you decided might be fine, but the way that you decided may make living with the outcomes a lot harder than it needs to be. If you've already made the decision just to get on with life, Chapter 5 – Decision Traps and Escape Hatches provides help in dealing with the later pitfalls of this kind of decision.

No one has specific information about what the future holds for you,* but you do have a lot of information about yourself, what decisions you've made in the past, how you came to them, which ones you found pleasant and which ones you found unpleasant to live with. This is the information that can help you make your own best decision to stay in or get out of the military.

> *To believe with certainty, we must begin by doubting.*
> — *Stanislaus*

1.1 – USING YOUR OWN LIFE AS A MAP

The first step is to take stock of what you already know about your own decision-making style. The reason for doing this is to avoid making a choice driven only by your decision-making habits of the past. You may not be aware of such habits, but your past decisions almost certainly were strongly shaped by them.

> *He who has begun has the work half done.*
> — *Horace*

To free you from these past patterns and let you focus on a clear, rational decision, the Lifemap was developed. This first chapter shows you how to use the Lifemap to capture your decision-making history quickly. Its visual format makes it easy to refer to, as you go through the four steps of making your decision about staying in or getting out of the military.

Fill out the Lifemap by following the examples and instructions in this chapter. This exercise will provide you with an important overview before you move on to lay out and compare your Options. (It's always useful to have the Big Picture in mind before making Big Decisions.) The Lifemap takes you on a tour of the decisions that led

---

* **Blind Choices:** "We...recognize the points along the road of our life where we came to intersections, and where we had to choose which road to follow. Very often, at those moments of choice the signposts were unclear. We had limited information available to us, and we had no way of knowing what sort of terrain lay further along the road we would choose. Very often, also, we had to make our decisions in the midst of the pressures of events, while traveling at full speed and without an opportunity to stop and study the alternate possibilities. Of necessity, then, we have all made many blind choices at the intersections of our life." [1]

to where you are now in your life. With this benefit of hindsight, you can look at the outcomes and how they impacted your life.

Some people recognize their decision-making patterns just from a quick, first pass through the Lifemap. Figure 1–1 is a Lifemap which was filled out in only ten minutes. That person then reflected on it and realized that he always found change painful, even when the outcome was fine. He was surprised to see that he likes slow, steady change and advancement. This, in turn, woke him up to the fact that it would take planning to achieve a second career that had a similar pattern of steady progress. Suddenly, he was acutely aware that he only had two years until his 20-year mark. Until this exercise, he'd been intending to wing it, to find a "retirement" job just to bring in money. He now realizes that he wants a position where he can keep getting promotions and has begun his search to find a serious second career with a clear advancement track.

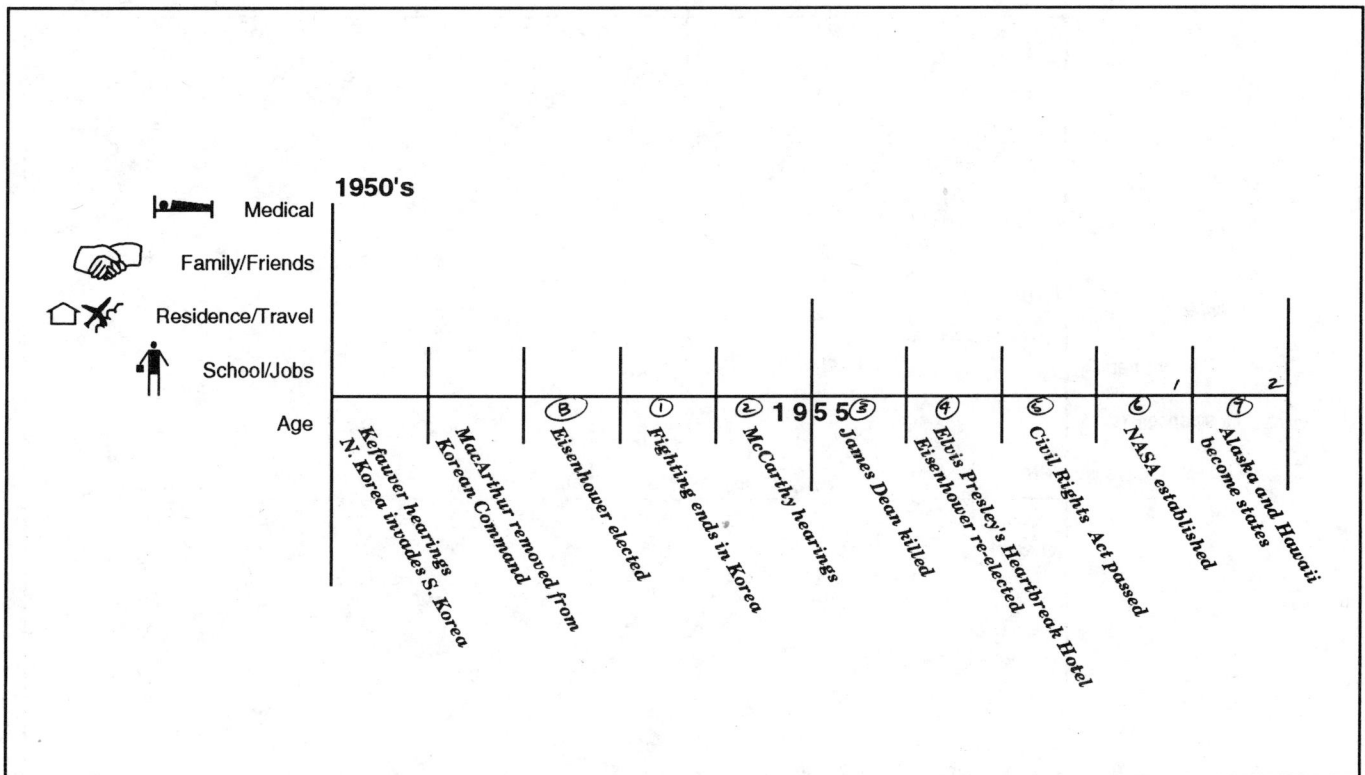

1950's

Medical
Family/Friends
Residence/Travel
School/Jobs
Age

⑧ Eisenhower elected
① Fighting ends in Korea
② **1955** McCarthy hearings
⑤ James Dean killed
④ Elvis Presley's Heartbreak Hotel / Eisenhower re-elected
⑤ Civil Rights Act passed
⑥ NASA established
⑦ Alaska and Hawaii become states

Kefauver hearings / N. Korea invades S. Korea
MacArthur removed from Korean Command

Figure 1–1: Example of Quickly Filled-Out Lifemap (Part 1)

1960's

Medical

Family/Friends

Residence/Travel — MOVED TO OAK ST. HOUSE →

School/Jobs

Age — 3 | 4 | 5 | 6 | **1965** 7 | 8 | 9 | 10 | 11 | 11

⑧ ⑨ ⑩ ⑪ ⑫ ⑬ ⑭ ⑮ ⑯ ⑰

- U-2 spy plane shot down
- Bay of Pigs invasion — Alan Shepard first American in space
- John Glenn first American to orbit Earth
- John F. Kennedy assassinated — Civil Rights March on Washington
- Gulf of Tonkin — Johnson re-elected
- Wide-spread anti-war protests — Power failure in northeast US
- Bombing of Hanoi
- White, Grissom & Chafee perish in NASA fire
- US Pueblo seized — M.L.King/R.F.Kennedy assassinated
- Anti-Vietnam War march on Washington

1970's

Medical

Family/Friends

Residence/Travel — → | MOVED OUT! | USAF ALA! | XFER CORPUS

School/Jobs — / | / | * | ↗12 | ↓ | ↓MARIA! | | +/ GAIL | ↓ !+ PROM'N | SCHOOLS →

Age — 18 | 19 | 20 | 21 | **1975** 22 | 23 | 24 | 25 | 26 | 27

INS. CLAIMS PROCESS

- Students killed at Kent State — Beatles disband
- Pentagon papers in NY Times
- Watergate break-in
- Viet Nam peace pact signed
- Nixon resigned, pardoned by President Ford
- US civilians evacuated from Saigon
- Rocky wins Oscar — US bicentennial celebrated
- Roots televised
- Panama Canal agreement signed
- Hostages taken in Iran — Three Mile Island

1980's

Medical — BACK PROBLEM | | | | | | BACK SURGERY +-↓↓

Family/Friends — | | MARRIED !+ | | DAUGHTER BORN !+

Residence/Travel — | | | | XFER E.COAST | | | XFER CALIF.

School/Jobs — SCHOOLS → | | ↗ +! PROM'N | ↓ ! PROM'N | | PROM'N! | | | INSTRUCTOR

Age — 28 | 29 | 30 | 31 | **1985** 32 | 33 | 34 | 35 | 36 | 37

- John Lennon killed — Reagan elected — Mt. St. Helens erupts
- Air traffic controllers on strike
- 1st American gets artificial heart
- 1st American woman in space — Begin seeking cause of AIDS
- US invades Grenada — 1st woman nominated VP
- Live Aid concert for African famine
- Challenger crew perishes
- Missile reduction treaty between US/USSR
- Bush elected
- Alaska oil spill — Berlin Wall down

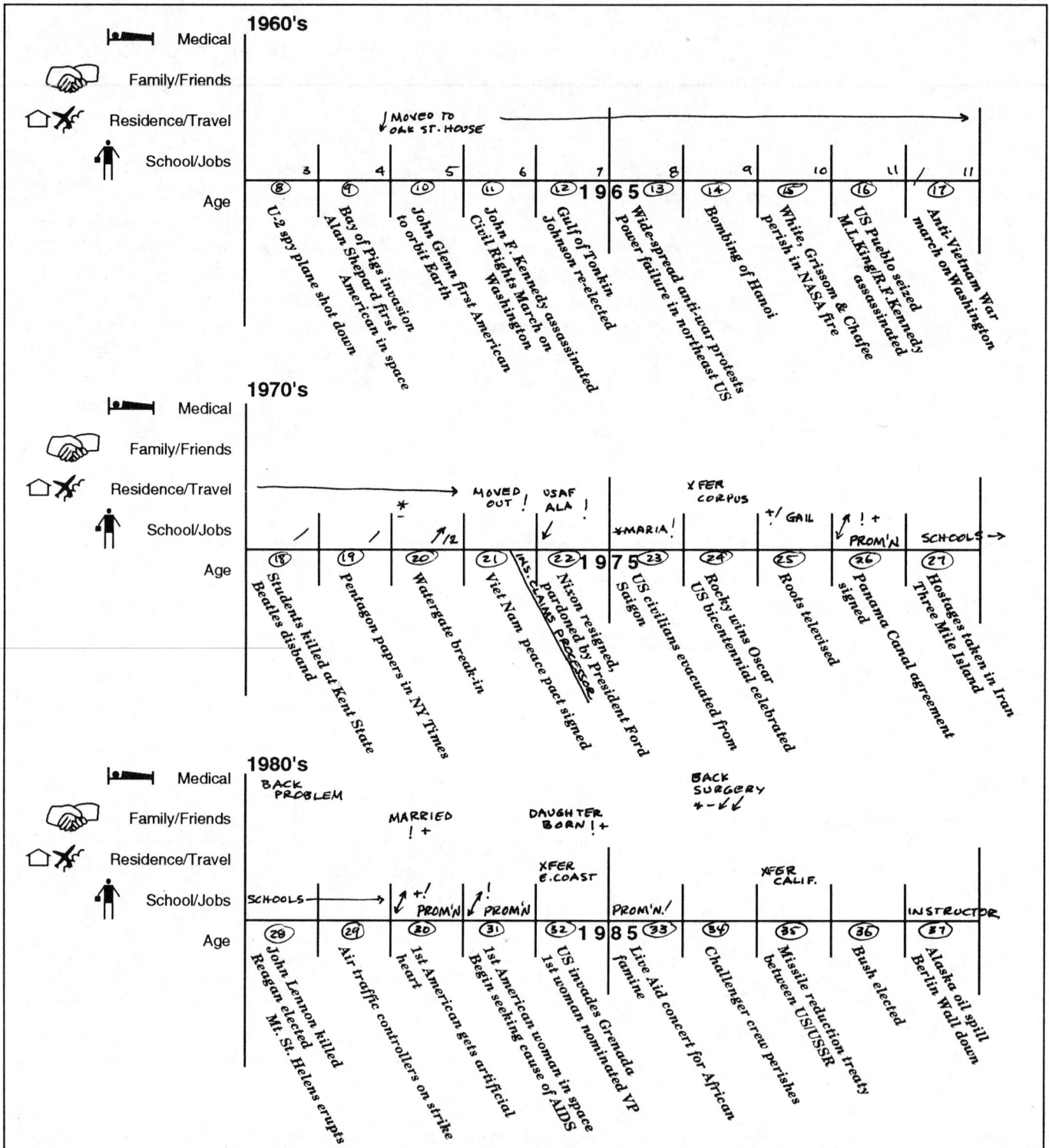

Figure 1–1: Example of Quickly Filled-Out Lifemap (Part 2)

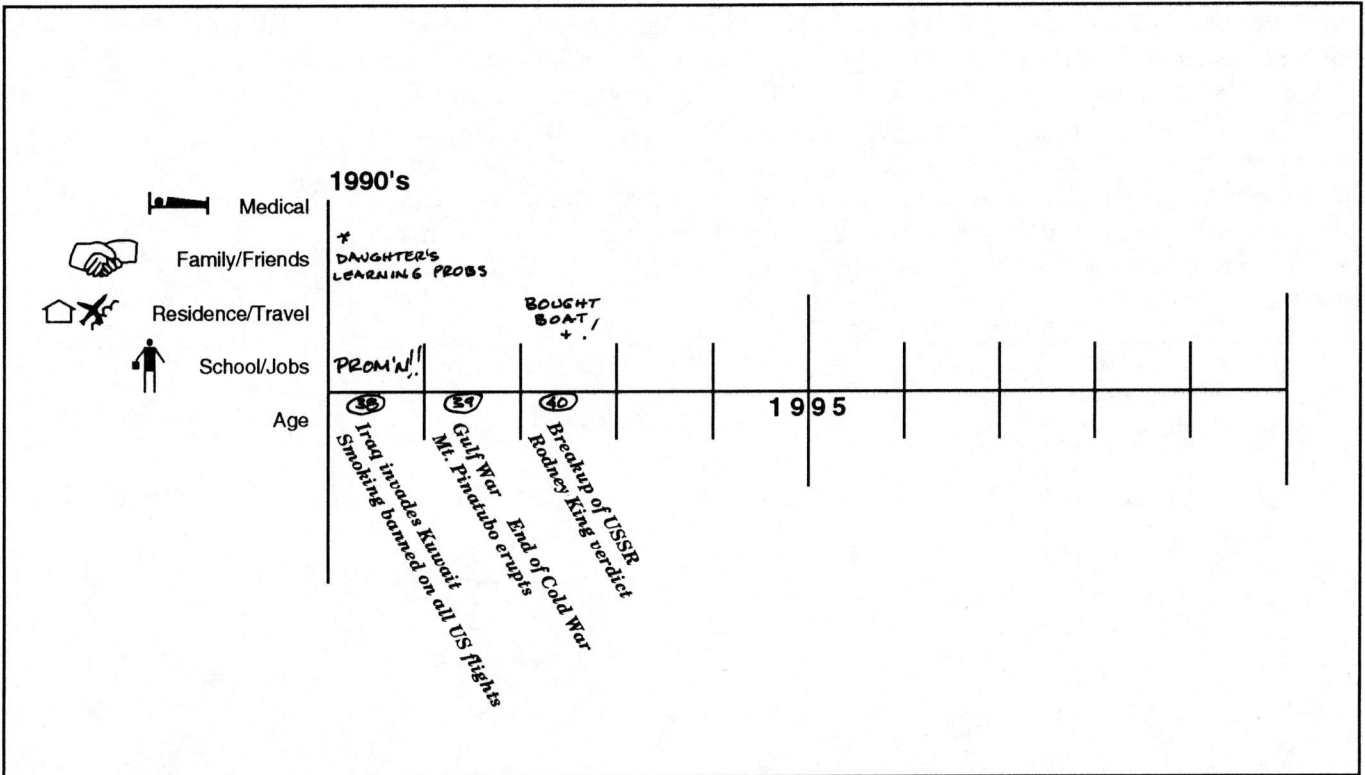

Figure 1–1: Example of Quickly Filled-Out Lifemap (Part 3)

Another benefit of filling out your Lifemap, whether quickly or in detail, is to look closely at how change and decisions have been connected to stress and challenge in your life. This first chapter helps you understand your personal patterns of handling change, so that you clearly see those factors that make change and decision making stressful and those that make it a challenge. From your Lifemap, you can more accurately predict how change and decision making will affect you when you face them again.

A second example of a Lifemap (Figure 1–2) was filled out more completely, in a little over one hour. The person then used the information on stress and challenge in the second part of this chapter to add more details, such as who made each decision, which were losses or gains, and which felt good or bad. From his detailed Lifemap, he realized two things that changed how he faced his decision. First, he saw that he hardly noted where he lived at different times, except for once. Second, he noticed he had negative reactions each time he was sent somewhere by the military, even though he often ended up being pleased with the new location. Taking these two issues together, he realized how much he missed living in Hawaii (the one residence he had noted as significant on his Lifemap) but that he'd given up on any more big moves because of the negative reactions associated with anticipating them. For the first time, he is seriously considering going back to Hawaii; he has begun discussions with his family and is getting his wife to fill out her own Lifemap.

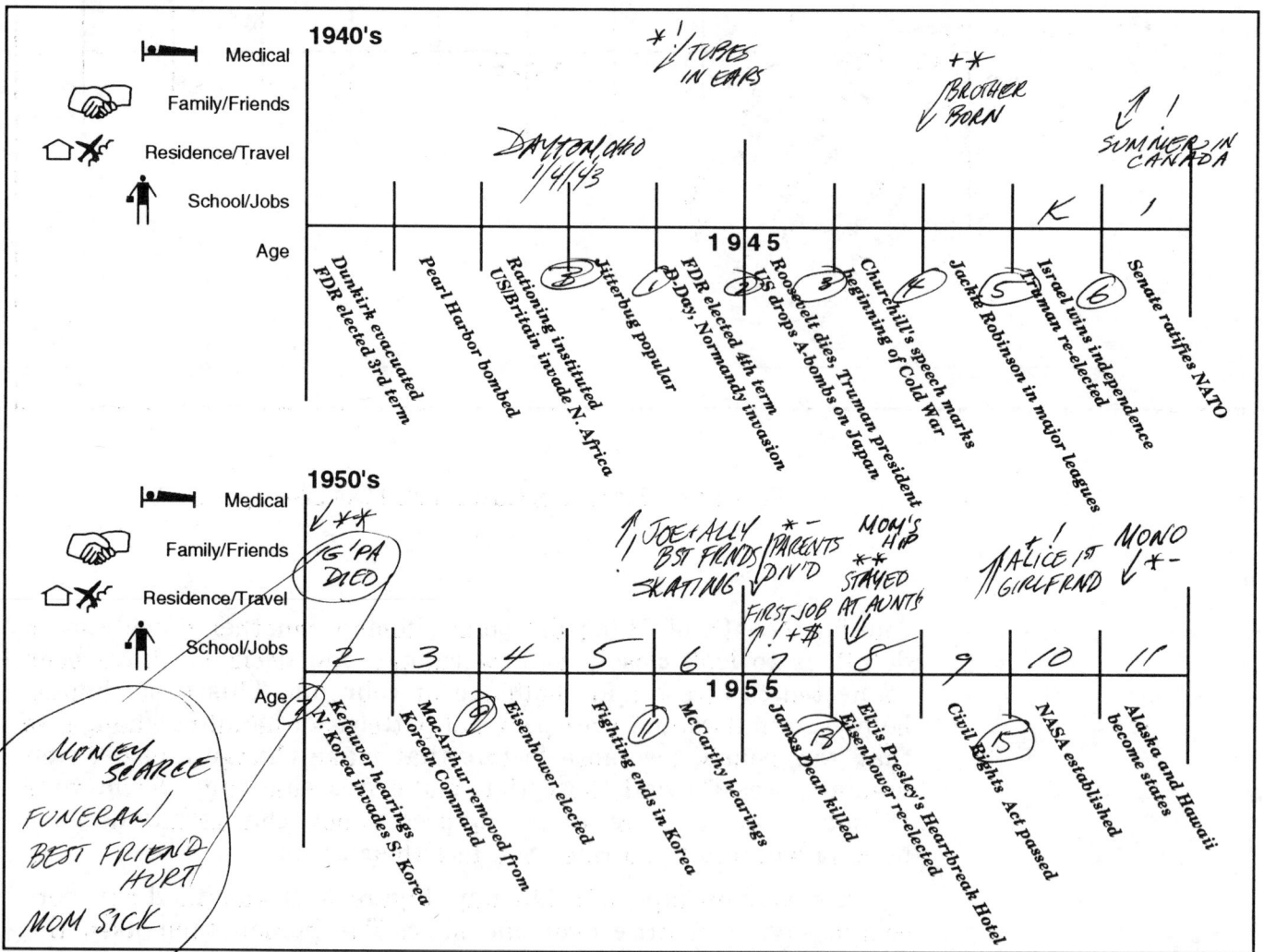

Figure 1–2: Example of a Lifemap Filled-Out In Detail (Part 1)

1960's

Medical
Family/Friends
Residence/Travel
School/Jobs

Age

| 17 | 19 | 20 | 21 | 22 | 23 | 25 |

FIRST CAR ↑+!
MET 1ST WIFE ↑+!
MARRIED ↑!+ / NAVY
NAVY CRUISE *-↓ PRO/+
SEPARATED ↓ -* / VCRS
DIVORCE ↑MET MARY
BROTHER MARRIED / CRUISE ↓-*
MOTHER DIED *-↓ / MARRIED MARY ↑+!
NAM LAST CRUISE
SON BORN ↑! / ↑+ OUT OF NAVY

- U-2 spy plane shot down
- Bay of Pigs invasion / Alan Shepard first American in space
- John Glenn first American to orbit Earth
- Civil Rights March on Washington
- John F. Kennedy assassinated / Johnson re-elected
- Gulf of Tonkin
- Roger failure in northeast US / Widespread anti-war protests
- Bombing of Hanoi
- White, Grissom & Chafee perish in NASA fire
- US Pueblo seized / M.L.King/R.F.Kennedy assassinated
- Anti-Vietnam War march on Washington

1 9 6 5

1970's

Medical
Family/Friends
Residence/Travel
School/Jobs

Age

| 27 | 28 | 30 | 31 | 33 | 35 | 36 |

SON'S EYE SURGERY *↓
RE-ENLISTED ↓ *-
PI TRANSFER ↓
STEPSON IN TROUBLE / NAVY SEC ↓
ON LEAVE !!++ / HAWAII ↓ !!+ -*
CRUISES ↓↓
MARY WORKING ↓ +

- Beatles disband / Students killed at Kent State
- Pentagon papers in NY Times
- Watergate break-in
- Viet Nam peace pact signed
- Nixon resigned. / pardoned by President Ford
- US citizens evacuated from Saigon
- Rocky wins Oscar / US bicentennial celebrated
- Roots televised
- Panama Canal agreement signed
- Hostages taken in Iran / Three Mile Island

1 9 7 5

1980's

Medical
Family/Friends
Residence/Travel
School/Jobs

Age

| 38 | 39 | 40 | 41 | 43 | 45 |

HEART PROB.

- Reagan elected / John Lennon killed / Mt. St. Helens erupts
- Air traffic controllers on strike
- 1st American gets artificial heart
- Begin seeking cause of AIDS / 1st American woman in space
- US invades Grenada / 1st woman nominated VP
- Live Aid concert for African famine
- Challenger crew perishes
- Missile reduction treaty between US/USSR
- Bush elected
- Alaska oil spill / Berlin Wall down

1 9 8 5

Figure 1–2: Example of a Lifemap Filled-Out In Detail (Part 2)

After you have reviewed the example Lifemaps to see how others have approached this process, prepare to complete this first step of your decision by photocopying blank Lifemap pages from the back of the book (Appendix C). Fill out your own Lifemap using the instructions that follow, and then read the rest of this chapter, keeping your Lifemap at hand.

☆ ☆ PHOTOCOPY LIFEMAP PAGES ☆ ☆

It is important to fill out your Lifemap, whether sketchily or completely, so that you can look for your past decision-making patterns. More importantly, you will benefit by adding to your filled-out Lifemap at a number of stages as you continue reading.

LIFEMAP INSTRUCTIONS

Each portion of the Lifemap has a timeline covering ten years. Under each line are some world events that were going on at that time, to help you remember your life events. The life events that you will be recording may be one-time occurrences (got first motorcycle) or the start of ongoing situations that are important in you life (met lifelong friend, chronic illness diagnosed, etc.).

1. **Age:** Begin by locating the timeline for the decade in which you were born and enter your birth "B" below the line. Now do the same for your age at every birthday following (1, 2, 3, etc.) using additional timelines to bring you to your current age.

2. **School:** Write in school years above the timeline (1, 2, 3, etc.). For college use Roman numerals (I, II, III, IV, etc.). Use M1, M2, etc. for military schools; G1, G2, etc. for graduate school years; and P1, P2, etc. for other professional training.

3. **Changes:** Your Lifemap should now have numbers both above and below the timelines. Use them as reference points for entering changes. You will first log the changes in your life. Then you will add marks showing how you reached those change points and how you felt about them.

    Mark all life changes by slash marks (/). Use the amount of detail you think will be useful for each period of your life, as follows:

    a. **Work:** Write in your job history using a slash mark (/) to indicate job changes. Note salary changes and promotions.

    b. **Residence:** Write in each major place of residence or assigned duty station. Summarize travel changes when significant (e.g. "back and forth, Phoenix and San Diego").

    c.   **Family and Friends:** Write in major relationships and their changes: births, marriages, deaths, beginnings and endings of friendships.

    d.   **Medical:** Write in both chronic and acute problems, noting changes that are important for you.

4.   **Type of Change:** Now go back to your slash marks and add arrow heads to indicate whether you chose the change for yourself (↑), others chose the change for you (↓), or the change was based on chance or mixed in choice (↕). You can add the feelings that accompanied changes: glad (!) or painful (∗), or whether the changes were an addition to your life (+) or a loss (−). Use space below the line to note rejected Options, the roads you did not take.

☆   ☆   FILL IN YOUR LIFEMAP   ☆   ☆

If there are events that you want to record in more detail, use the zoom lens technique where an entry is circled and then expanded in the margin for a close-up look (as shown on the 1950 line of Lifemap Figure 1–2).

1.2 – CHANGE AS CHALLENGE, CHANGE AS STRESS

Now that you have your life changes summarized on your Lifemap, it's time to review the changes you've made in your life. Any questions you have about where to put different items will be answered as we go along. If there are problems you can't get past, consult the examples to see how other people used the Lifemap to lay out their histories. Remember, it's a tool for organizing your own particular experiences in making and weathering the changes in your life, so there's no right or wrong way to use your Lifemap.

Understanding how you have made changes in the past is important in understanding your own personal patterns, especially around changes you have initiated and decisions about staying or leaving.

A second and more basic reason for focusing on change in your life is to discover when change results in stress and when it leads to challenge. One of the factors that keeps change challenging rather than stressful is the degree to which you initiate the change. Usually, the more say you have in a change, the less stressful it will be.

1.2 A – WHO CONTROLS CHANGE?

When you made your first entries in your Lifemap, you indicated places of change by slash marks (/). Most of these are at forks in the road, times when you began or ended a phase of your life.* Review how much control you had over these changes, noticing: (1) which ones you initiated (↑), (2) which ones were strongly influenced by circumstances or other people (↓), and (3) which ones were a mixture of both or chance (↕). The mixed ones include those that you can't remember well and those where many prior small decisions fed into larger decisions.

There's no right or wrong to this, it's how much you feel that it was you who made the change versus some outside force. From this review, notice whether one type of decision is most typical for you. Also notice whether the control of changes shifts over time.

☆ ☆ REVIEW YOUR LIFEMAP ☆ ☆

1.2 B – AMOUNT OF CHANGE

Too much or too little change can both be sources of stress. We've all had jobs that had so little change** that they were plain boring and other jobs where events happened so fast that there was stress from sheer exhaustion.

The concept that there is an amount of change that is just-enough is depicted in the U-shape curve shown in Figure 1–3. The left side shows that too little change results in boredom and stress, while moving too far to the right results in stress from overstimulation, anxiety, and disorientation. The just-enough point is somewhere in the middle. It is different for each person and may shift for the same person at different times in her or his life. A good name for this is the "Goldilocks" function, where everybody wants to find the amount of change that is not "too little" or "too much" but "just right."

---

* **Roads Taken:** "Some...changes took place because of decisions which we ourselves made. Others took place because of decisions that were made by others, or that were forced upon us by the impersonal circumstances of life. However they occurred, they were particular experiences which determined the direction and shaped the contents of our lives from that point onwards." [2]

** **Tiresomeness:** "...a...work situation sometimes becomes so tiresome or grinding that he feels compelled to think about quitting. His distress becomes a vital force, pushing him to risk more than he might ordinarily dare. Thus long before new goals are formulated, discomfort can serve as an impetus for change." [3]

For military folk, the experience of too much and too little change is common. The too-much change of sudden moves and the too-little change of marking time often go together in a one-two combination of too much followed by way too little change ("Hurry up and wait!"). Much of the early research on stress emphasized the negative effects of too much change, and gave little attention to the effects of too little change.

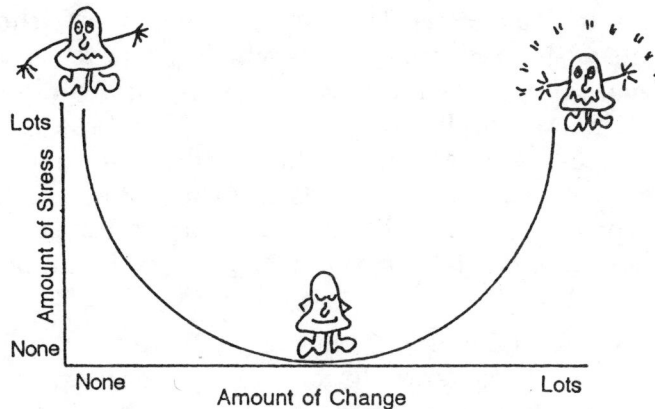

Figure 1–3: Stress as a Function of Change

Another important factor is how central the changes are in your life. Two major change areas for military folk are geographical moves and shifts in the amount of authority you have. Some geographical moves trigger many other changes. For example, the cruises of Navy personnel are not just geographical shifts, but bring many differences in danger, working conditions, and family life in their wake. A geographical move, where the whole family relocates to a different part of the country but the active duty person's job is essentially the same, also requires major but different kinds of adjustments from a cruise or deployment.

A shift in the amount of authority deserves special emphasis because promotions and demotions affect how you interact with other people all day, every day. Usually, authority shifts result in a change in how you see yourself as well. An increase in supervisory responsibility changes your working day, how you act with co-workers, and, in many military situations, the people you hang out with in your social life as well.

Keeping in mind that larger changes may involve geographical moves or shifts in degree of authority, review your Lifemap and emphasize with double slash marks (/ /) some of the larger changes in your life.

☆ ☆ REVIEW YOUR LIFEMAP ☆ ☆

1.2 C – PAINFUL AND PLEASANT CHANGE

Some theories of stress propose that the amount of change is so important that it even overrides how you feel about the change. Research suggests there is equal stress whether you feel positive and eagerly anticipate an event or you feel negative and would avoid it if you could. In this research, stress is defined as "any change that places an extra demand on you." The results showed that equal stress results from positive as well as from negative changes when measured by the frequency of illness two years after the changes.[4] Based on these findings, just facing the change itself can be a barrier to leaving the military.

Mark those life changes on your Lifemap that were especially painful and especially pleasurable for you. Use the asterisk symbol for pain (*) and an exclamation point for changes that felt good (!).

☆ ☆ REVIEW YOUR LIFEMAP ☆ ☆

1.2 D – ADDITIONS AND SUBTRACTIONS

A final way to view changes is as additions to your life (birth of a child) or subtractions (best buddy moves away). Pay attention to this difference because there are emotional side effects of subtractions that will come up as you contemplate leaving the military. Whenever people or experiences leave your life or you leave them, there is a separation, a disconnection.

Leaving anything can be tough. Even when you leave a place that you just can't wait to get out of, there's often a moment of doubt, wondering whether you'll be sorry or not, a moment of reluctance to make it final. People hesitate to trade in their old car, even when the facts say it's the smart thing to do. Remember, you can end up missing a toothache or even a mole that you always wanted removed.

Social psychologists suggest that there are two parts of leavetaking: the pain of the separation itself, and the losses once you've left. Take, for example, a man who left home long ago and now hears that his father has died. There is pain from the separation by death, but in many ways he had separated when he left home. He feels this pain of the final separation during the funeral. In addition, there is the loss of the role of being a son. Whatever this man did as a son with his dad is over now. Visits, calls, letters, and just thinking how his dad would want to hear about his new car are parts of his role as a son which is now gone, or subtracted from his life. Leavetaking includes both (1) the pain of separation from familiar people and experiences, and (2) the loss of events, relationships, and roles that help define and shape your life.

Reviewing your own past decisions to leave is a way to understand what you're facing now. What has it been like when you've decided to end a relationship — to leave that woman or that particular guy? What relationships have you ended yourself? Have you left jobs? Houses? Towns? Under what conditions? What did you call it? Bailing out? Cutting your losses? Getting free? All are leavetakings to be added to your Lifemap.

Some people avoid leaving because of the echoes of pain from being left. Being "dumped" is especially memorable when it's sudden or messy. Afterwards you may vow "I'll never do that to anyone." Depending on how vehemently you make such promises, they can affect your decision making for a long time.

We don't need to get too philosophical here, but every time you make a choice, it requires leaving some other options behind. Once you pick one direction, you usually have to give up the possible opportunities that were in the other directions. You may wish to note these directions-not-taken on your Lifemap as well. Take your decision to enlist as an example. What other options were you considering at that time? One major one? Several vague ones? Directions that you wanted to avoid anyway? You can mark these unchosen options below the timeline on your Lifemap. Figure 1–1 shows the Lifemap of a person who gave up a 9-to-5 job processing insurance forms when he enlisted.

Perhaps as important are the times you planned on leaving someone or something but didn't. What happened? Did you honestly reconsider or did you just run out of energy? How did you explain to yourself those times when you made the decision to leave but then didn't? If you have faced the decision of leaving the military before, note this particular instance of avoided change on your Lifemap as a "road not taken."

These three types of experiences — the times you left, the times you were left, and the times you planned to, but didn't — are your "leaving" history, your on-the-job training for making the decision you are facing now. Add them all to your Lifemap. Mark the subtractions and leavetakings with a minus sign (–). Write in any planned endings that never happened below the timeline.

☆ ☆ REVIEW YOUR LIFEMAP ☆ ☆

You may be saying that all this emphasis on leaving is unbalanced because (a) you may stay in and (b) you're not planning to say goodbye to anyone until you have a solid plan for what you're going to do next.

Of course, you're right.

You'll want a clear idea of what's next before leaving anything — a clear idea of the new jobs, locations, and people you are heading toward. It's not helpful to emphasize leaving over arriving, but leaving comes first in time and needs to be faced first. Even if leaving causes only the loss of stuff you've wanted to be done with, there is still the pull of the familiar, the nostalgia for predictable patterns. The loss of the predictable way an awful boss starts a meeting is still a loss. What might be coming is not as clear as what you're leaving behind. If you stay in but make a major change in your military career, you're also facing many goodbyes. Add other losses that come to mind to your Lifemap as minuses. Look over the whole pattern of changes on the Lifemap. Make notes of any consistencies or surprises.

☆ ☆ REVIEW YOUR LIFEMAP ☆ ☆

If you are in a hurry to carry out your decision steps **and** if you feel that stress is not a factor for you, skip the rest of this chapter and go on to Chapter 2 – Step 2: Laying Out Your Options. If you have been stressed about your current job or about making your decision, complete this chapter.

1.3 – KINDS OF STRESS

The Lifemap is useful for getting an overall picture of the changes in your life so far and how you dealt with them. With this general idea of your history of changes, let's look at how some changes become stressful and some become challenges.

General stress management theory says that all changes are potentially stressful.[5] Each change that requires some adaptation on your part is called a "stressor." Whether you actually feel stressed or not depends on the amount and type of stressors you encounter (this

> *The trouble with the rat race is that even if you win, you're still a rat.*
> *— Lily Tomlin*

Section — Kinds of Stress) and the motivation, predictability, and control available to convert these stressors into a challenge (Section 1.4 — Stressors into Challenge).

1.3 A – STRESSOR LEVELS

Figure 1–4 is helpful to show how stressors work.

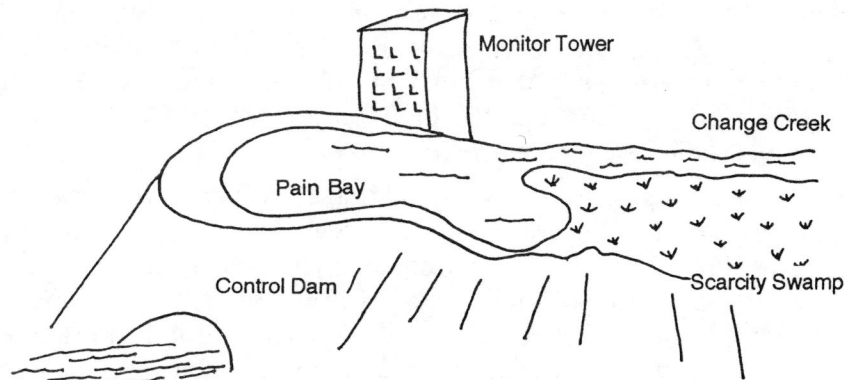

Figure 1–4: The Stress Landscape

Look at Pain Bay. The amount of water in the bay is based on the number and severity of stressors in your life. These in turn are based on the amount of change, pain, and scarcity that you have been experiencing. You can think of the stressors as filling up the bay; to avoid overflow, you can either reduce the amount of change, pain, or scarcity, or get a better handle on predicting and controlling them, depicted here as the Monitor Tower and Control Dam. Increased predictability and control allow people to withstand many stressors without major health problems. Stress management courses and books present a variety of coping tools to get better control over how you personally respond to stressors (see Appendix A).

1.3 B – CHANGE CREEK

You reviewed the role of change in your life when you filled out your Lifemap. Now you need to think about the amount of change in your current military situation. Is there too much or too little change? Would there be less stress in your life if there were less or more change? How much control do you have over the amount of change in your job and living conditions? Have you moved to totally different cultures and lived on the economy? Did you experience major changes when you moved back? Especially note how many of

The ultimate measure of a man is not where he stands in moments of comfort, but where he stands at times of challenge and controversy
— M. L. King, Jr.

these occurred during the last two years, since those in that time period are likely to still affect your current stress level.

Review your Lifemap and note again where extra amounts of change occurred with double slash marks (/ /).

☆ ☆ REVIEW YOUR LIFEMAP ☆ ☆

1.3 C – PAIN BAY

You've already noted changes on your Lifemap that were especially painful or pleasant. Anything that results in physical or psychological pain goes into Pain Bay. Chronic pain is important. Also include physically painful environmental stressors such as chronic noise or lousy temperature control.

Social pain is also a factor: separation from those you love, being passed over for promotion, feeling forced to sit through boring stories, and living with political pressures that you'll never be able to beat, can all be painful. Your reactions may be different depending on whether there are chronic low levels or sudden unexplained events; even rumors of painful conditions can generate stress.

> *In this company,*
> *it is an unwritten policy*
> *that you work together,*
> *but burn out alone.*
> *— Veninga*

Included in Pain Bay are the threats and possibilities of major injury and death that can be part of any military assignment. This risk of physical pain or death is generally greater than in civilian life and, as a main source of stress, often goes unrecognized. Some of you have learned to live with this, tucked away in the back of your head, so that you are unaware of any effect on your daily functioning. Living with the threat of injury and death can be a constant energy drain or it may provide the extra spice that keeps you feeling alive.

You might need to adjust the pain events on your Lifemap as you think of pain in these ways.

☆ ☆ REVIEW YOUR LIFEMAP ☆ ☆

1.3 D – SCARCITY SWAMP

Scarcity includes both loss and having to do without. When you noted subtractions on your Lifemap, these included losses of various people and places, being left, and turning away from possible options (roads not taken). Here we look at why loss and scarcity may lead to stress.

Think of yourself as having needs and think of the environment, including the people in the environment, as supplying those needs.

Scarcity of supplies in this context can be a chronic state ("I never get enough!") or a loss ("What do you mean I can't go there anymore?") Supplies that are lacking can be either resources or reinforcers. Resources are those things you need to function. Depending on your lifestyle they would include anything from the necessities of life to the staff and budget you need to get a job done. Reinforcers are all those things that make you feel good about yourself, about how you're doing. They include recognition and approval, money, status, and experiences and objects you enjoy.

In the military, scarcity is related to both security and adventure in a unique way. The services provide a mix of high security in some areas right along with a high potential for risk and adventure. If risk and adventure suit you, you may not notice the scarcity involved, but now that you are considering leaving, it may be useful to know what conditions of scarcity do and don't bother you.

You can probably think of your military career as fluctuating between times of security and adventure. There are people who have been in for 15 years or more and had nothing but the security benefits of life in the service. There are folks who have been in for that length of time who have had high adventure and risk, year in and year out.

Think about security and its absence, scarcity, as you review your military pay, health care, housing, and training. Compare quiet times to those of risk and adventure, moving on short notice, living conditions when in action.

No matter what kinds of scarcities you've had in the military, any scarcity of resources and reinforcers was not your personal problem. This deserves to be underlined: it might have been your job to obtain scarce resources, but it was not your personal problem. Compare this situation to your pre-military days when scarcity of resources was solely your responsibility and realize that whenever you become a civilian again you'll probably be directly responsible for resources again.

Revise and adjust the times of scarcity and loss on your Lifemap.

☆ ☆ **REVIEW YOUR LIFEMAP** ☆ ☆

1.4 – STRESSORS INTO CHALLENGE

There are four other major aspects that affect whether a stressor will actually stress you or be converted into a challenge: (1) the motivation factor, (2) the degree of predictability, (3) the degree of control, and (4) the degree of reversibility of an event.

1.4 A – MOTIVATION AND SENSE OF MISSION

Caring about what you are doing, what you are spending your time on is one of the best ways to neutralize stressors. The degree of importance and commitment versus triviality that you bring to a situation provides a kind of motivation control. If you are clear about why you're doing what you're doing, if you believe that your efforts contribute to a worthwhile mission, then the impact of many stressors will be lessened.* This is not a book on motivation or on career planning in general, but without motivation to accomplish goals, you have very little protection against the impact of those stressors called change, pain, and loss. Your motivation may be as simple as a goal of staying on a job so you can pay bills; if you're clear about it, it's a fully worthy goal and a committed attitude can neutralize the stress of many otherwise unmeaningful jobs.

Caring about what you're doing is one way to keep stressors from turning into stress. Another way is to deny stressors. Denial of stressors by focusing on optimism helps in situations where you have done all you can do and the rest is out of your control. This occurs in medical situations like pre-surgery, where, once you have hired the best surgeon you can to do what you yourself cannot do, denying your fears can increase calmness, hope, and faith which are helpful for healing.[6]

Since it's unlikely you can hire a specialist in deciding when you should leave the military, it is unlikely that denial will help in this case. Only you can do the homework in this book, lay out your military and civilian options for the next few years, compare them and decide.

> The first and worst of all frauds is to cheat oneself.
> — Philip James Bailey

1.4 B – PREDICTABILITY

Being able to predict stressors will reduce the stress you feel from them. The Monitor Tower at the edge of Pain Bay depicts this. If you have no control over a stressor, but good predictability, it will cause less stress than otherwise. If an uncontrollable but predictable change is coming and it is positive, the predictability·turns into anticipation and reduces the stress.

---

* **The Alexander Syndrome:** This syndrome is named after the guy who ran out of lands to conquer. "In our society, lack of continuing progress feels like failure, especially to men in their middle years who have become accustomed to constant climbing as a major source of gratification. Trained to get somewhere, they feel defeated when they finally arrive." [7]

If the stressor is expected to be painful or a loss, there are additional ways to reduce the stress.* First, you may be able to find meaning, a sense of mission, in putting up with the stressor. If you can see its value for you, for the people you care about, or for values you hold important, then even an uncontrollable but predicted negative change can be neutralized by accepting it. Once accepted as necessary, you can free yourself to be objective about what aspects will be most painful for you and get help with those rather than toughing it out pretending it's not painful. Navy couples use this approach when the men and women who go off on long cruises believe that their work is important, but do not deny to their spouses the difficulty of being apart. Facing the parts that are toughest allows people to do some creative planning for those left behind and creative packing for the person leaving.

If you can't find any meaning in a predictable but uncontrollable painful change, you may turn to the "I'll show 'em!" denial pattern. The mental focus of this pattern is on authority figures, people who directly control the stressors or powers-that-be who decide from on-high. A mental dialogue goes on with a jaw-gritting style ("Don't let the turkeys get you down."). This pattern has a sense of personal mission, a call to perform in spite of uncontrollable and meaningless changes and demands that are totally predictable. Righteous anger can be harnessed to keep going in spite of the system.

> *We first make our habits, and then our habits make us.*
> *— John Dryden*

Such a pattern based on dogged endurance can help people stay in a situation they'd rather leave and often results in high stress levels. Personality types who struggle angrily against adversaries and authorities often find it difficult to change to any other pattern.

Other personality types may benefit from a group sharing of frustration. When the whole group draws together and says, "Don't let the turkeys get you down," the camaraderie and shared survival can make it an intensely positive time. People in battle situations against a common enemy have this intense closeness and unity. The pattern is exactly the same even if the perceived "enemy" is the person in charge. Contrast battle situations with bureaucratic ones: in the paperwork bunkers of office situations, people rarely can find a common enemy since each one is struggling to maintain his or her

---

* **Hardiness:** Using a Life Change Survey, researchers looked at executives with high stressor scores to see which ones subsequently became sick and which did not. The executives who did not become sick were (1) highly committed to what they were doing, (2) felt that they controlled their own lives rather than being controlled by circumstances, and that they (3) looked for new experiences in their lives rather than sticking with the secure and familiar. In their tendency to search out novelty, these "hardy" executives were not foolhardy; they took calculated risks rather than risking just for the sake of adventure.[8]

own job or competing for the next promotion. Bureaucratic politics rarely reinforce people working together.

1.4 C – THE CONTROL DAM

Look again at the Stress Landscape in Figure 1–4. Notice the Control Dam. Having control of change, pain, or scarcity lowers the probability that these stressors will result in actual stress; having control is one of the main ways of converting change into challenge instead of stress. Learning to run the Control Dam changes the threatened flood of stress into the power of challenge. There are two different aspects of controlling change that can be useful for reducing stress. There is the preference for control and then there is the perception of control.

Research shows that people choose the more controllable option when they can, even if they never actually exercise control but have only the potential of control. It even occurs when kids are asked to choose, or have someone else choose, one of two identical cans to get what is hidden inside. Most kids prefer to make their own choices even when, as in this situation, it's a blind choice.[9]

Other research emphasizes that people vary in their desire for degree of control and like to be in control of how much control they have! This theory points out that people fluctuate in their preferences and only they have access to how they are feeling on any given occasion. If they have a range of controllability, they can adjust the amount used to match their internal state of the moment.[10]

It is possible that in your situation, where you are faced with a life-change decision, it's the moment-to-moment shifts in how much control you want that keep you on a seesaw. One minute you get excited about a plan for your future, not wanting input from anyone else; the next moment, you'd pay someone else to decide for you.

Although the research generally agrees that people like to choose their own changes, it gives mixed answers to the question of why this is so. Kids are not the only ones who play their hunches, perceiving that they have control when logic would say they don't.

There are also studies of people who routinely perceive themselves as having little control. "Learned helplessness" is "the belief that one's actions are independent of consequences — a lack of personal control." Evidence that such beliefs are associated with clinical depression has led to treatment programs that emotionally inoculate depressed people against helplessness. Such programs include training in how to decide which situations are controllable and which are not, learning to guide your actions with positive self-talk, and a wide range of other mental and behavioral coping skills.

You will face factors you cannot control as you make your decision about staying in or leaving the military. Learning to assess how controllable they are and watching for the natural tendency to get depressed about them can be like a vaccine to inoculate you against learned helplessness.*

Sometimes, of course, people may not be overperceiving their helplessness, but are accurately assessing there is nothing they can do to affect the outcome. While people who are prone to depression don't differentiate controllable from uncontrollable factors, overly optimistic types may believe they have control when, in fact, they don't. They have exaggerated the influence they have on outcomes, and increase their risks of being blindsided by just those factors they mistakenly thought they could control. They may not have to struggle against feeling hopeless or depressed but may have to suffer surprises and shocks when events don't turn out as they expected.[11]

You used the symbol ↑ to note when you controlled a change and ↓ when you didn't. Review and revise these events on your Lifemap as you read on.

☆ ☆ REVIEW YOUR LIFEMAP ☆ ☆

1.4 D – REVERSIBILITY AND WORST-CASE SCENARIOS

One of the hitches in having control over a decision is that once you've made a decision, you may forget that it's reversible. You, yourself, may, in the future, tell yourself, "That's it! You chose it! Now you live with it!" as if you can never change your mind, bail out, or cut your losses. Certainly if you feel you are locked into any situation with no options left, you cannot feel very much in control of your life nor exert much control over the stressors that occur. It doesn't matter if you are the one who carefully designed and built the corner where you feel trapped. What matters is that you feel you have made an irreversible decision.

You probably don't know how reversible your decision to leave the military will be. Many people assume that the decision to leave will be irreversible, that the military will not have you back once you muster out, that the door swings one way only, so that you can decide when to leave but you cannot reopen the door.

You need to research your situation. Even if the goals of the

---

* **Learned Helplessness:** Laboratory demonstrations with dogs show that after an initial stage where they receive a shock no matter what they do, they are unable to learn to avoid the shock when the means are provided. The animals end up cowering when they could jump over a low barrier to escape; they have been taught to be helpless. [12]

military are to reduce their numbers, you need to check if there are conditions under which the service would welcome you back. Age obviously plays a part, but also your skills, changes in technology, and national policy are influences that can close or re-open the door.

If you find that there are conditions under which you can return, you still have to guard against the tendency to box yourself into a corner by making your decision mentally irreversible. If you think, "Oh no! I'd never go back even if I could," that word "never" suggests an outcome you want to avoid. Fully facing anything you're avoiding helps focus your Options. Just asking yourself what's behind what you want to avoid can be useful.*

Now that you know four ways to change stressors into challenges, you may be realizing that you have a lot of stress going on. You might want to take some side-steps to reduce stress before you get on with Step 2 of making your In or Out Decision. Stress management can help a lot and help quickly, so the sooner you begin, the better off you'll be as you face making your decision.

Sometimes the stress overflow is so overwhelming that an immediate time-out is called for. Taking leave time may be just what you need to find the distance from everyday concerns to get on with stress reduction and decision-making steps. Appendix A lists some of my preferred books on stress management; you may find it easier to get your stress levels in hand if you get someone to help you.** Counselors, stress management classes and workshops, or mental health professionals can help you lay out a stress management plan to implement in your daily life.

If most of your stress is from the part of you that wants to get on with your decision to stay in or leave the military, you will not want to delay; move right on to Step 2 – Laying Out Your Options.

---

* What Are You Avoiding: Asking yourself, "And then?" after each answer brings you closer to understanding what you're wanting to avoid. For example, "If I went back in the military, I'd have to start at the bottom." And then? "And I'm not going to do that because I'd come in at a much lower rate than I left with." And then? "People who knew me before would look down on me and think I couldn't make it on the outside." And then? "Well, I'd be embarrassed and humiliated and slink around for the rest of my time." And then? "OK, OK, I'd either keep doing that or get tired of it and get on with my life."

** A Short Commercial: If you feel very stressed, you will want to consider professional counseling especially if you can see that depression is a factor and has been for a while. Stress management is not a substitute for therapy and a course of psychotherapy may be what's needed if depression, substance abuse, or chronic anxiety have entered the picture as they often do around mid-life and career decision times.

CHAPTER 2

STEP 2 — LAYING OUT YOUR OPTIONS

Laying out your Options is trickier than it seems. Step 2 can lead in two directions. Either it will go easily, so you can move right on to Steps 3 and 4, or you will run into a problem and need to use the problem-solving help in Chapter 5 before going on to Steps 3 and 4.

So go ahead and start. You will find out soon enough where any problems are.

2.1 – HOW MANY OPTIONS?

A major way to avoid confusion among your Options is to have more than two. This will help keep your personal seesaw from going back and forth forever with no satisfactory decision.

So, instead of having two options —

OPTION 1 Stay in Military	OPTION 2 Re-Enter Civilian Life

you would have —

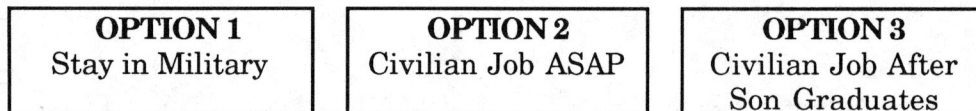

OPTION 1 Stay in Military	OPTION 2 Civilian Job ASAP	OPTION 3 Civilian Job After Son Graduates

or —

OPTION 1 Stay In, Keep Status Quo	OPTION 2 Stay In, Change Job	OPTION 3 Re-Enter Civilian Life, with Job ASAP

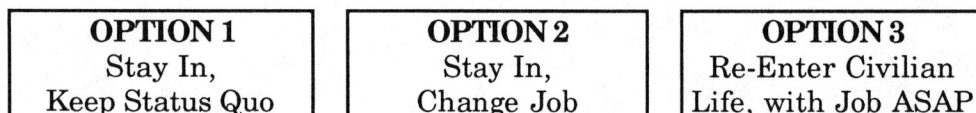

There are many reasons to have more than two Options. First, if you have any trouble at all making up your mind, broadening your range of Options is likely to give you more room to work out a decision that fits you. Having only two Options prematurely narrows your Options. Practically everyone can come up with at least two Options for staying in and two for getting out. If you are in the situation of retiring from the military and not planning to work afterward, having more than two Options of when and how to retire can help you restructure your overall goals.

Most active duty personnel have several Options to change their role in the military if they stay in. Include as many Options for change as possible as you begin to lay out your Options. Techniques for narrowing your choices and weeding out Options that turn out to be unacceptable will be detailed in Step 4.

Beyond the general guideline of starting out with more rather than fewer Options, research suggests that the more Options you have, the more you will feel in control of your life. In one study, for example, "a moderately large number of Options (six) was associated with greater perceived choice than was either a small (three) or a very large (twelve) number of Options..." The higher number of alternatives was associated with feelings of choice and enjoyment of the activity.[13] Be sure to put in an Option for any lifelong dreams you want to head toward.

Decision making gets more risky when you have only two Options. Figure 2–1 shows you in the middle between two Options. You are getting dizzy; you are polarized by your Options.

Figure 2–1: Polarized by Two Options

Polarization results from having only two Options, each with positive features (PROs) and negative features (CONs). This situation is also called an approach-avoid conflict.* In such a conflict, as you get closer to a decision (approach), the importance and weight of various doubts increases and you want to back off to avoid them. In a way, it's like the changes in visual perspective that occur as you move through space.

Figure 2–2 is a diagram of this changing perspective.

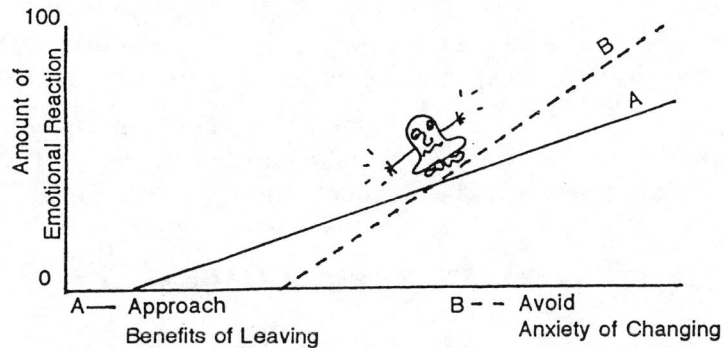

Figure 2–2: What a Way to Build a Seesaw!

As you first approach the idea of re-entry into civilian life, for example, you may focus on the military frustrations you will get away from, the potential benefits of leaving (A). But then, as you get psychologically closer to leaving, you begin to feel the natural anxieties that go with change: doubt ("what if I can't get a job?"),

---

* **Approach/Avoid Theories:** Approach/avoid behavior implies that anxiety about going is stronger than the frustration of staying. It also implies that the anxiety increases more rapidly than the frustration as we approach each Option.[14]

increased heart rate, stomach flutter, etc. You begin to avoid and back away from the anxiety of going, also backing away from a decision to leave (B). Then, as you move back toward the idea of staying in the military, you re-emphasize the frustrations of staying until they loom up and push you toward leaving. This is one psychological explanation of the back-and-forth quality of indecisiveness.

If you feel stretched out tight between two alternatives, you may already be so polarized that you find it hard to think of additional Options. One way to come up with more Options is to target different times for leaving the military. These time points might be based on when your pension and benefits change or on when you expect other events in your life to occur, such as when the house sells, when savings reach a certain amount, or as soon as a certain type of training is completed.*

Basing Options on different time points can solve the problem of being polarized, but has one small hitch you have to keep an eye on. The hitch is that the Option farthest out in time can seem less likely just because it seems so far away, less real, than those Options that are looming up closer. Also, the further benefits and costs may also seem less real (see Figure 2.3).

> *The lure of the distant is deceptive. The great opportunity is where you are.*
> — John Burroughs

We all know about the problems that can come from minimizing the negatives that are way off in the future (buy now, pay later; eat now, diet later). We can also needlessly close the door to good opportunities by not keeping the far-away benefits clear in our minds.

Go ahead now and rough out your choices as you currently see them in the first column of the Table of Options (page 27).

☆ ☆ FILL IN YOUR OPTIONS ☆ ☆

Once you've laid out more than two Options, you can work with them, refining them to be most useful to your particular situation.

---

* Retirement Date Options: "I am somewhat surprised at the small number of people I have encountered who took advantage of the flexibility of timing that is so frequently possible in planning retirement. Most people start with their "official" retirement date and work backward in their planning, trying to arrange all their affairs to culminate on the target date, and often never seriously consider the possibility of shifting the target to a more advantageous date. Mandatory and official retirement dates are made to seem so immovable as to make any thought of change fruitless. Yet, on closer examination, one sees that in most plans, a change of retirement date is not impossible; it almost always requires paying some sort of financial penalty, but sometimes it will make better sense to take the penalty rather than remain pinned to an arbitrary date. And sometimes the penalty is not great." [15]

TABLE OF OPTIONS

OPTIONS	AMOUNT OF CHANGE	DEGREE OF PREDICTABILITY	AMOUNT OF LEAVETAKING
1.			
2.			
3.			
4.			

Make sure each has a time frame. Specific times are critical because they will affect the types of information you'll need about each Option. Also, if your Option times change, you'll know you need to update information as well.

2.2 – WHAT KIND OF OPTIONS?

2.2 A – DEGREE OF CHANGE

Look at each Option in terms of amount of change required. The goal is not to quantify the amount of change exactly, but to see the relative amount of change required among your Options. Some Options may call for return to skills or a lifestyle already familiar to you, while others may require starting from scratch in various ways. It will require more change for you to run a small farm if you've never identified a plant in your life than if you have the necessary skills. After all the talk about change leading to either challenge or stress, you can understand why it's useful to estimate the total amount of change each Option will require.

☆ ☆ FILL IN CHANGE ESTIMATES ☆ ☆

Fill in your entries in your Table of Options for amount of change using descriptive words (lots, little, too much), or by numerically ranking the Options (#1 – most change, #4 – least change).

As you estimate amount of change for each Option, you will also realize that the less predictability you have for an Option, the shakier are your estimates of change.

2.2 B – PREDICTABILITY AND ABILITY TO CONTROL

The further away in time an Option is, the less predictable it seems. But there are some exceptions. For example, re-entry into a well-known job, or a situation where you will have a lot of control over what you do, may have high predictability no matter how far in the future it falls.

Many aspects of civilian life make it less predictable than the military. For starters, once you are a civilian, you can be more easily fired. You become more responsible for life decisions in a different way than in the military, having to work out your own ways of handling things like health insurance and retirement planning. If you have made this independent decision to leave, it will only be the first of many.

> *Habit is a cable; we weave a thread of it every day, and at last we cannot break it.*
> *— Horace Mann*

Your military Options may also range in predictability from a solid stairway of predictable promotions to pie-in-the-sky possibilities that depend on making unusual rank changes. Now that enlistment, promotion, and retention levels are being shaped by international politics, changes can occur rapidly and without warning. Forecast which skills and billets should be stable and which may change with national defense priority shifts.

☆ ☆ FILL IN PREDICTABILITY ESTIMATES ☆ ☆

Mark (using pencil) each Option for level of predictability. You may also wish to reword an Option, revise it, or add or delete whole Options.

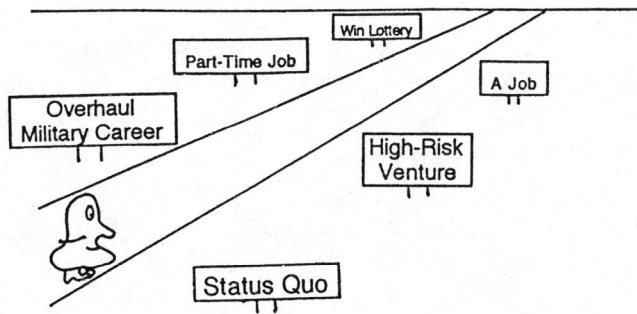

Figure 2–3: Time Perspective

2.2 C – LEAVING

Finally, each Option needs to be looked at in terms of the amount of leavetaking involved. You've reviewed other times that you've left when you filled out your Lifemap in Chapter 1. Now your job is to examine each Option in terms of the relative amount of leaving. Consider not only people and places you'll leave, but also the goals you have set, the unfinished business that you won't be able to complete now. Maybe there's something as minor as getting a section of files overhauled that never got done because it could always be put off until tomorrow.

There are many aspects of leaving that can be painful. As you consider leaving a particular niche, you may run up against suspicions and questions about who will miss you and about how essential you were in the first place. This is a natural time to look back over your career or stint, reviewing the high and low spots, jobs done and left undone.

Many social settings try to ease the leaving process with ceremonies, parties, and rituals to help people finish up and provide closure. Saying goodbye, parties, and other rituals of leaving can emphasize or ease the strain of leaving. The better you know yourself, the better you can forecast the effects that leaving and its "celebrations" will have on you.*

Mark your Table of Options for the total amount of leavetaking you expect to occur for each Option. Although those Options that are far off in time are harder to predict, your goal is to pin down the relative amount of leavetaking among the Options as best you can. Adjust and revise your Options as you think of better wording.

☆ ☆ FILL IN LEAVETAKING ESTIMATES ☆ ☆

By now your Table of Options is well marked up with flags, arrows and changes. You may realize that you need more or fewer Options or that you need more information about them before you can do any more comparing.

Your Options may be very different from one another so that it feels like comparing apples and oranges, or, more dramatically, like comparing apples, motorcycles, beaches, and diesel fuel. Such dramatic differences are easier to keep in mind. When your Options are different only along one dimension, they are less dramatic, like comparing apples of different sizes; less dramatic but simpler overall.

You may be ready to copy out a clear revised Table of Options or you may want to do more research on your Options first. Go ahead. Get started on a list of tasks and phone calls that will improve your Option information. Following some basics about PROs and CONs that complete this chapter, you will get on with balancing your PROs and CONs in Chapter 3 – Step 3.

2.3 – PROs AND CONs

Almost everyone has approached decisions by taking paper and pencil to list the PROs for and CONs against an available Option and

---

* **Endings:** "...without a proper ceremony of closing and renunciation, they have no way to mourn the cases that now become forever unfinished business, to honor the intentions that will never be fulfilled. To turn away from them so casually is to trivialize them and to diminish themselves." [16]

the anticipated outcomes that go with it. It is this quality of being "for" and "against" various outcomes that can make you feel like you are on a seesaw. This section will help you follow through with that natural impulse in ways suited to deciding whether or when to leave the military.

Making a list of PROs and CONs is a good start, but it's just a start. For example, say you are only considering whether or not to buy a small boat that someone has offered you a good deal on. You write:

PRO   CON

and then make all the entries you can think of. Figure 2–4 shows that, in deciding whether or not to buy a sailboat, our character thinks that the chance to sail in the sun is a PRO while the risk of being out in a storm is a CON. Your PROs and CONs, of course, may be very different.

If you're facing a small decision, such a list may help, and, in fact, be all you need. For a big decision, the amount of information can quickly get overwhelming. Some people give up on such lists even for small decisions because the lists seem like velcro, attracting more information than you need. Imagine that you keep adding to the list about the boat, for example, until you can't think of anything else. Perhaps you carry out the process over several days, including information from phone calls and conversations with others (e.g. the moorage rates at the marina, which lakes have loading ramps, current price of fuel, the cost of foul weather gear, etc.). When the list begins to get sidetracked into trying to plan your family vacations for the next five years to justify the purchase, you are in trouble.

Figure 2–4: PRO/CON Decision — To Buy a Boat or Not

However, if the list is brought back to target, it can be helpful. When you research a particular boat thoroughly, the risk of unwelcome surprises down the road is kept low. Usually, decisions about purchasing a standard product, one that has been made and used before, present minimal risk, and that risk can be lessened by talking to other owners. Vacation packages, entertainment, and other experiences have more risk since people generally can't compare experiences as easily as products. Keep this in mind when you listen to other people describe their career change experiences.

If you treat your decision to stay in or leave the military as a single PRO/CON decision, you are probably oversimplifying such a big decision.

2.3 A – DEGREE OF RISK

Decisions vary in their amount of risk because of differing amounts of unobtainable information. Some decisions are commitments over so long a time span that there's no way anyone can predict all the relevant variables (marriage, jobs). Other decisions involve situations that are actually chosen for their unpredictability (sports, investments, gambling).

The unpredictability or risk may stem from lengthy time spans, unknown factors, or risk built in for fun and excitement. Figure 2–5 describes decisions ranging from low risk to high risk.

10% Risk	50% Risk	90% Risk
Trades, custom-ordered items, bonds, purchase of items in stock	Purchase of first-time items, real estate, sports games, marriage	Speculative investments, purchase of used and auction items, business partnerships

Figure 2–5: Degrees of Risk

Typically, in a PRO/CON decision for starting something new, you know more about the CON side, since you are comparing something new to the known status quo and the new always involves some risks. In the decision of whether to stay in or leave the military, you know more about staying than going. In general, there is less known risk in sticking to the status quo than in sticking your neck out. The risks of not changing tend to be less visible and sneak up on you over time.

2.3 B – PRO/PRO CHOICES

Leaving the military is much more complicated than purchasing a boat because if you say "No" to the boat, you simply don't have a boat. If you say "No" to leaving the military and decide to stay in, you still have to deal with whatever got you thinking about leaving in the first place. For example, if what started you thinking about leaving was the supervision style of your senior officer, what do you do about continuing to work for him or her if you stay?

This type of CON against staying can become a PRO for getting out. As you list your PROs for and CONs against staying in or leaving the military, any particular anticipated outcome may end up as a PRO on one side or a CON on the other. The effects of phrasing items as negative or positive are important and this shows up as a psychological effect of whether we state an item as a PRO or as a CON. Technically, this is called *framing*, how we frame each predicted outcome.* The frame of each item will greatly influence your negotiations with yourself.

You also need to notice that where factors end up, as PROs or as CONs, depends on what you compare them to. For example, if the military has been subsidizing your family's travel for years, getting a job with an airline with travel benefits can be seen as a PRO for cheaper travel compared to what other civilians pay or as a CON for travel compared to what you paid in the military.

Whoops! Now we've got two kinds of lists. We've got PROs and CONs on one list, but also comparisons of the PROs of leaving with the PROs of staying. This second list looks like:

PRO: leaving   PRO: staying

You are now looking at a choice between two alternatives viewed positively. I am using "choice" to refer to decisions viewed in this PRO/PRO format because "choice" typically implies selecting (1) freely from among alternatives that (2) have a high degree of predictable outcomes that are (3) positively viewed and (4) carry no risk of awful outcomes.

---

* **Reframing Negotiations:** "You bought your house in 1982 for $60,000. It is now on the market for $109,000, with a real target of $100,000, which you estimate is the true market value. You receive an offer of $90,000. Does this represent a $30,000 gain (compared to the original price) or a $10,000 loss (compared to your target price)?

"Negotiators need to be aware of how framing affects the decision process. If you are evaluating a negotiation in terms of what you can lose, make sure you also consider what you can gain, and vice versa. Otherwise, your behavior may reflect the distortion of framing rather than your actual preference for a particular action.

"...we confirmed the finding that people who frame the outcomes of negotiation in terms of gains or profit are more willing to make concessions to obtain the sure outcome available in a negotiated settlement. In contrast, negotiators who think in terms of losses or costs are more likely to take the risk-seeking action of holding out and possibly losing all in an attempt to force further concessions from their opponent."[17]

Combination Plate #1   Combination Plate #5

PRO   PRO

Figure 2–6: PRO/PRO Choices — Which Chinese Dinner to Buy

People make such free "choices" among PRO/PRO Options when there are no restrictions, threats of punishment or negative consequences. An example of a PRO/PRO comparison is when you're choosing what to eat from a menu and there are several things that you like, and money and calories pose no limitations. Without writing lists down, you mentally compare the Options, as shown in Figure 2–6. Unlike a PRO/CON, which has a CON that "you can't refuse" to pay attention to, PRO/PRO decisions are ones that you really can refuse; you are free not to choose at all.

Ever since the Middle Ages, philosophers have been trying to predict the "logical" thing to do in such choice situations. They used the example of a donkey standing halfway between two identical bales of hay. Later, the existential philosophers stated that one has to take a leap of faith, just choose.

One of the nice things about PRO/PRO choices is that there is usually less anxiety about making them than making decisions where strong CONs are looming. No matter which you choose, you're going to like it. For those of you who are positive about the military and positive about a civilian future, your main job will be maximizing the positives. If you choose to re-enter civilian life, you will probably have some negatives about the change itself but that is different than predicting negative long-run outcomes. Review your Table of Options to further highlight your feelings about leaving, before continuing to read about the PROs and CONs of your future Options.

☆ ☆ REVIEW TABLE OF OPTIONS ☆ ☆

Most purchase decisions are PRO/PRO choices; they are relatively risk-free and there are no big risks of punishment or unpredicted aspects. You often can tell how much risk people perceive in the words they use. Purchasing an item with most of the information up front is typically called a choice; without much information, a gamble. If the gamble turns out well, it's a bargain; if not, and we want to save face, we call it a high-risk investment.

Now you have seen two types of lists you can make. If you're considering *whether or not* to do one action, a PRO/CON comparison will help. If you're considering two positive Options,* but they are mutually exclusive in that you have to choose one and reject the other, a PRO/PRO comparison is helpful. You use both types of thinking in purchasing decisions. First there is the PRO/CON decision whether or not to buy a particular type of item. Then, if you decide to buy, there is the PRO/PRO decision of which item to buy.

You could also look at a decision in terms of only its negatives** by making a CON/CON list of anticipated outcomes, but the anxiety of only focusing on negatives is likely to increase decision-making stress drastically. If you tend to focus on CONs, that's fine, but read on to see how to include some PROs in your overall view.

2.3 C – DOUBLE PRO/CON CHOICES

Now let's focus on deciding whether to stay in or leave the military. Earlier we talked about each Option as really two decisions that are back-to-back — staying in the military and postponing civilian life or leaving the military and re-entering civilian life. You're going to get a chance now to use both the PRO/CON sheets and the PRO/PRO sheets, this time together. We begin by building a double PRO/CON comparison that includes the PROs and CONs for staying in as well as the PROs and CONs for leaving the military. Selecting one Option always involves some PROs and some CONs, even if it

---

* **Freedom:** Generally we think of freedom as present in those situations where there is no threat, no coercion. Skinner alerts us to dignity as well as freedom. Generally he describes freedom as people doing what they choose in spite of high risk of punishment while dignity is people following their own ways in the absence of any reward for it. He notes, "...the literature of freedom has been designed to 'make men conscious' of aversive control (e.g. punishment) but that ... it has failed to rescue the happy slave (controlled by rewards)." [18]

** **CON/CON Comparisons — Your Money or Your Life:** Happily, there are not many career decisions that involve comparisons of outcomes that you would only be against (CON vs. CON). In the short run, of course, everyone has felt that they were in a situation "between the devil and the deep blue sea"; in some sense, you might call that a kind of freedom, but in terms of career decisions, being stuck between two equally negative Options is a clear signal to get help in generating more Options!

only means giving up the PROs and CONs of the rejected Options.*

On the PRO side of either Option, there are things that you are glad to gain and also things you're glad to get out of. On the CON side of either Option, there are things you are unhappy to have to put up with and things that you are unhappy to lose. A sample double PRO/CON comparison is shown in Figure 2–7.

Figure 2–7: A Double PRO/CON Decision — Comparing the PROs and CONs of Two Options at Once

* **Choosing among Mutually Exclusive Options - From the Diary of a Georgia Philosopher:** "With age, I'm realizing that each time I decide to do something, I'm also deciding not to do more and more other things. It used to be, when I was 11, a choice of shooting baskets out back or going on down to the drug store to see who was there. Now, if I go play tennis, I have to give up watching a tape, reading, balancing the checkbook, washing the car, returning my neighbor's ratchet wrench, and/or having a beer...the Saturday afternoon options are endless. I don't mind doing any one of them; it's just hard to give up the others." [19]

2.4 – OUTCOME FRAMES

2.4 A – WORDING OF PROs AND CONs

Noticing how you use PROs and CONs in comparing Options is a way of using decision framing to improve your decision. With practice, you can get quicker at phrasing items as PRO or CON, and, as you read on, you will see how the way that you frame your PROs and CONs can be useful.

To review, comparisons of various Options will feel different depending on whether you frame your concerns in terms of the PROs or in terms of the CONs.* Let's say losing access to military gyms is a concern for you. This item might show up as a PRO for staying in or as a CON against re-entering civilian life or both. Trust your instincts and put such entries wherever they make the most sense to you. Don't worry about having the same item occur in several places on your Table of Options. Which events you consider as PROs and CONs will be unique to you and your life experiences. Only you can tell whether health concerns for you or your family or the availability of travel work out to be a PRO or CON for you. The reason to trust your instincts is that whether you say that an item is a PRO for staying or a CON for leaving will affect how you weigh these PROs and CONs later.

You might think that an item should **weigh** the same whether it's a PRO for staying or a CON for going. Not true. Research says that we tend to use different weightings for comparing PROs than those we use for comparing CONs. Notice:

"Polls show that 51% of the public feel that the President is handling the hostage crisis well." Well, good.

"Polls show that 49% of the public disapprove of the President's handing of the hostage crisis." Uh-oh!

The rule of thumb for keeping your perspective on gain or loss relative to some comparison item is, *people will take a sure gain and avoid a sure loss.* This rule was developed by studying how people compare gains and losses in many kinds of situations. In your case, the gains are the PROs in your predicted outcomes and the losses are the CONs.

---

* The Law of PRO and CON Relativity: "Thibaut and Kelley emphasize that it is not the 'absolute' amount of gain and loss he expects to encounter that determines the value a person will place on a given choice, but the amount relative to a comparison level, based on the amount of reward or punishment the person has obtained in the past or has seen other people obtaining." [20]

Let's apply the rule of thumb that people take a sure gain and avoid a sure loss to the situation of whether to stay in or to leave the military. Add in your common sense because the research was not done about such big life decisions, and few of the outcomes you'll be predicting will be absolute 100% sure-things or 0% no-chance-at-alls.

In comparing the gains of staying in versus getting out, the gains of staying are going to look better to you because you are surer of them.* You will tend to want to take the sure gains of staying in, because any gains of getting out have more risk associated with them. When comparing gains, the average person would choose the sure gains of staying in. When comparing losses, the picture leads to a different conclusion. Comparing the losses of staying versus getting out, you will tend to avoid the sure losses or CONs that got you considering leaving in the first place, and choose the uncertain losses of a change.

Just by the nature of the way people compare gains and losses of goodies, you can see that the deck is stacked in favor of your staying in until the gains or losses waiting in the civilian life become as predictable as what you already know about the military. In addition to the gains and losses of the goodies that might go with a job and lifestyle, there are also items you have-to-put-up-with or are glad-to-get-out-of.

Let's say you'd be glad-to-get-out-of military paperwork and for some reason you cannot change your job in the military. You already know how much paperwork there is in your current situation; that's

---

* Take a Bargain, Avoid the Sure Loss: Let's say you have a choice presented to you framed in terms of gains: you can choose an 85% chance to win $1000 (with a 15% chance to win nothing) or the alternative of receiving $800 for sure. Most folks take the $800 sure thing. Now we have to look at this rule of thumb very closely to see how it improves plain common sense. What you are doing is taking a sure gain of some money (100% probability of $800) over a risky gain of more money (85% probability of $1000). Notice the overall odds are better for the risky gain:

Option #1:	$	1000	x	0.85 (probability)	=	$	850
	$	0	x	0.15 (probability)	=	$	0
Option #2:	$	800	x	1.00 (probability)	=	$	800
	$	0	x	0.00 (probability)	=	$	0

In other words, choosing between a sure gain of $800 and an on-the-average gain of $850, people choose the sure gain, which is the lower average gain.

But look what happens when these same alternatives are framed as losses. Consider the situation where an individual must choose between an 85% chance of losing $1000 (which is also a 15% chance to lose nothing) and a sure loss of $800. People avoid the sure loss, choosing an 85% chance of losing the $1000, which is the higher average loss:

Option #1:	$	−1000	x	0.85 (probability)	=	$	− 850
	$	− 00	x	0.15 (probability)	=	$	−0
Option #2:	$	−800	x	1.00 (probability)	=	$	− 800
	$	− 00	x	0.00 (probability)	=	$	−0

a sure CON for you. There's no way you can predict, however, how much paperwork there will be in your civilian role. Suppose you guess that if you get out, there's a 15% chance of doing all paperwork (and an 85% chance of doing none) or a sure bet on continuing doing the 15% paperwork if you stay in. The rule of thumb says you probably would choose the chance to get out of all of it (85% of no paperwork) to avoid the sure loss (100% chance of 15% paperwork).

Overall, then, we can apply the rule of thumb about sure gains and losses to your decision about the military as follows:

1) Assume you know more about the PROs and CONs of staying in than you do about re-entering civilian life. You know more about the gains and losses of staying and so they are "surer" than those of leaving.

2) Assume that these surer gains and losses around your decision operate in the same direction as "sure" ones in the psychological research.

3) Notice that the rule of thumb would lead to different choices depending on whether you focus on PROs or CONs. In general, you would:

   a. If considering CONs, decide to re-enter civilian life to avoid the surer losses of the known military life, and

   b. If considering PROs, decide to stay in the military to take the surer gains of the known military life.

Thus, if this general rule to avoid sure loss and take sure gain applies to your situation, there are several things to look out for:

- Emphasizing CONs over PROs in your thinking leads to bias toward re-entering civilian life

- Emphasizing PROs over CONs in your thinking leads to a bias toward staying in.

2.4 B – COSTS VERSUS LOSSES

One last idea before you use this knowledge about framing PROs and CONs. Items that get framed as CONs can be looked at more closely. Some make a distinction between CONs that are costs and those that are losses.

Costs usually refer to predictable amounts of money or time you trade for benefits. Losses are more generally thought of as something that you have and then have to give up for no agreed-upon benefit. An example is how people frame insurance premiums. They can be framed as a *sure loss* (paying the premium) chosen over the *risk of a loss* (an accident), or as the *sure cost* of protection against the *risk of cost* of an accident. Generally, people react more negatively to losses than to costs. Framing CONs as losses or costs helps explain choices that otherwise do not make sense.

Consider a man who develops tennis elbow after he's already paid a hefty club membership fee. If he continues playing, he's considering the fee as a cost of playing tennis; having paid the cost, he'll play in spite of the pain to gain the benefit he paid a cost for. If he stops playing, he's considering the fee as a loss. Sometimes, we continue something we'd rather not, just because we already paid for it and are unwilling to change it over to a loss in our minds. We paid a cost for "it" and we will collect "it" even if it is no longer pleasurable, useful, or if collecting "it" causes great pain.[21]

Each of the above ways of framing an event will influence whether you phrase it as a PRO or a CON. You may need to refer back to this section as you lay out the anticipated outcomes for each of your Options in terms of PROs and CONs.

CHAPTER 3

STEP 3 — TRIAL BALANCE

In this third Step, you will compile the anticipated outcomes for each of your Options on a Trial Balance Sheet (page 43). Using what you know about framing the outcomes, you'll list them as PROs or CONs. Now, enter your Options, in pencil, across the top of the Trial Balance Sheet so that there is a column for each Option. If you want to redefine an Option at any time, do it. (An extra copy of the Trial Balance Sheet is provided in Appendix C.)

☆ FILL IN OPTIONS ON YOUR TRIAL BALANCE SHEET ☆

Keep in mind the ideas you've read about the effects of framing outcomes as PROs or CONs and begin to enter Anticipated Outcomes as either PROs or CONs on your Trial Balance Sheet.

In addition to the Options written in across the top, a checklist of ideas for the anticipated outcomes for each Option is entered down the left-hand side. Four sections of anticipated outcomes have been set up to help you differentiate those that are your concerns from those that are concerns of other people, and to separate anticipated outcomes that are related to practical matters from those concerning self-image and approval.

The items can be practical and objective or feeling-oriented and subjective. Your pencil-only habit means you can just write down whatever enters your head the first time through without worrying where an item fits best. For example, having an interesting level of work might seem to be an item that impacts you more than others. In some families, though, it might be also important to anyone tired of listening to your complaints of boredom.

These categories are suggested only to help you think through anticipated outcomes, not because there are clear black-and-white distinctions among them. For example, the freedom to work on your own may be a practical item for you because you know for a fact you work better without lots of supervision. Or it may be a subjective item because you prefer less supervision, even though you can work well under either set of conditions.

The entries on your Trial Balance Sheet may include a wide variety of items. Janis and Mann[22] have used this type of sheet with people who are making a variety of career decisions. The categories, which have been adapted for military personnel's career decisions, are reviewed in Section 3.1. The order of the categories within each of the four sections generally moves from those that enhance security to those that enhance risk and independence.

You may choose to first fill out the Trial Balance Sheet before reading any further. If you do it this way, it will be clearer which items are spontaneously important to you. If you choose to do this, go ahead and think about your first Option. Spend some time imagining how it would look, feel, and sound to be living out that particular life. Then set a timer for five minutes, relax by taking a slow deep breath, and begin to write in the anticipated outcomes that either please you or displease you. Enter these as PROs and CONs under this Option quite quickly, knowing that you can always erase and move them later. Just keep on writing whatever comes into your head, putting your entries into the spot that seems most likely to you at the moment.

Alternatively, you may want to read on using the categories in the Trial Balance Sheet when they relate to clear PROs or CONs for your Options. You can just make a check mark under PRO or CON for each category, adding notes when you want to remember details. To review from the last chapter, the PROs are gains or things that you're glad to get out of, while the CONs are the losses and unpleasant things that you'll have to do. You can use short phrases like "No, son's soccer games" and "Yes, short commute," or use plusses and minuses like "− son's soccer game" and "+ short commute."

TRIAL BALANCE SHEET

O P T I O N S

ANTICIPATED OUTCOMES	1. PRO	1. CON	2. PRO	2. CON	3. PRO	3. CON	4. PRO	4. CON
1. Practical Outcomes for You								
A. Income								
B. Job security								
C. Difficulty of work								
D. Stress								
E. Interest level of work								
F. Chance to use old skills								
G. Chance for advancement or learning new skills								
H. Room for making own decisions								
I. Chance to fulfill goals outside of work								
J.								
2. Self Approval/Disapproval								
A. Hating quitters								
B. Meaning of the work								
C. Links to long-term goals								
D. Need to compromise								
E. Creativity of work								
F.								
3. Practical Outcomes for Others/Family								
A. Income								
B. Status								
C. Time for family and others								
D. Decision making stress								
E.								
4. Others' Approval/Disapproval								
A. Prior commitments								
B. Status								
C. Independence vs. recognition								
D.								

To maximize the usefulness of this Trial Balance Sheet, you will need to start a list of "Information Needed." You'll be unable to rate many of the areas under each Option as PRO or CON until you go out and research more information about fields of work, training programs, living conditions and other aspects that are unclear to you. Here is one of the ways you can involve your family, if they've been wanting to help; send them on research treks to the library to check lists of facts and current almanac information that you need.

Each branch of the service has resource people to help with checking out the educational, financial, and other aspects of your military and civilian Options. Classes and workshops are available to assist in planning and adjusting to career changes, mid-life career choices, and retirement. Career and retirement counselors provide one-on-one sessions to review the relative benefits of each of your civilian or military Options. Be sure to seek their knowledge about the effects of timing on your Options. Vocational counselors at many local colleges can provide access to lots of information for free. Counselors who specialize in second careers may be well worth their fees.

Notice whether your counselors are in the military or not. You may need to correct your course to adjust for their biases, ignoring some of their enthusiasm or pessimism. Be sure to double-check all information you are given. This is your decision, not theirs. If you decide on the basis of information that is either biased or incorrect, you'll pay the price, not them!

Call or write old friends who have information about military programs and civilian businesses and jobs. If you decide to leave the military, there are many resources that can systematically guide you in transition planning.

Enter your anticipated PRO and CON outcomes even though you will be mixing those that are of great importance to you with those that matter very little. In your final summary sheet you will get to sort out those that weigh the most heavily with you.

☆ **FILL IN OUTCOMES ON TRIAL BALANCE SHEET** ☆

3.1 – PRACTICAL OUTCOMES FOR YOU

3.1 A – INCOME

Consider immediate pay level but also overall expected averages and probable rate of increase. Note situations where there is an opportunity for commissions, bonuses, or other benefits based on performance. These may be either a PRO or CON for you.

3.1 B – JOB SECURITY

Some Options can be judged directly in terms of job security. Jobs with the civil service, for example, are known as hard-to-be-fired-from. Although the military is usually considered to be secure, continued progress in a particular job classification is still tied to performance.

In some situations, job security is not directly predictable but can be figured out by deduction. In general, the more formal the evaluation and review system, the better your protection against suddenly being fired. Industries that are linked to fluctuating production schedules may include a high probability of both layoff and overtime. There's no substitute for talking to people already in the business. There are places with very formal evaluation systems that are window dressing only and everyone who works there knows that politics is the road to job security. Finally, there are some industries that are more likely to swing wildly with the changing fortunes of the economic marketplace than others.*

3.1 C – DIFFICULTY OF WORK

Physical hardship and long hours are certainly nothing new for military personnel. On the other hand, by the time you're thinking of leaving the military, you may well have moved up to better conditions in this area and it may be harder to readjust to any entry level "tough" conditions than you expect. If you're approaching 20 or 25 years of service, you should factor in the extra difficulty of adjusting to physically rough conditions.

Starting out in a new career is more difficult when you can't be sure what is expected of you. Look at each Option in terms of likely clarity of guidelines, amount of clarity of guidelines and structures typical of that industry and typical of that particular organization.

> *You won't skid if you stay in a rut.*
> *— Frank McKinney Hubbard*

---

* **Characterizing Careers by Risk and Feedback Speed:** The amount of money and time invested before you find out if you'll get ahead is one aspect of all industries. The speed of this feedback from the marketplace is another aspect that affects all careers in a particular industry.

High risk x	Quick feedback	Sports Heroes
High risk x	Slow feedback	Drug Researchers
Low risk x	Quick feedback	Mail-order Entrepreneurs
Low risk x	Slow feedback	Bureaucrats [23]

3.1 D – STRESS

The earlier discussions of stress can now be used to assess the amount that is likely for each of your Options. Look carefully for these factors in each of your Options: frequent reorganization (Change Creek), difficulty of the work (Pain Bay), and an atmosphere of time pressure and sudden deadlines (Scarcity Swamp).

3.1 E – INTEREST LEVEL OF WORK

Many classic mid-life career books ask people to figure out whether they are best at working with things and materials, with people, or with information. Other approaches ask you to decide whether you are interested in a particular content area (working with metal regardless of whether you design with it, solder it, or keep an inventory of it), or a particular function area (design, sales, or problem solving, regardless of the content). Your interest in each of your Options may well reflect either the content area or function area involved.

On your Lifemap, recheck any of your Roads Not Taken to discover old interest areas that you never followed up on. Factor in the energy of these old dreams that affects the interest level of each Option.

3.1 F – CHANCE TO USE OLD SKILLS

In the military, you probably have acquired many skills you can transfer to the civilian sector. You also have the experience of working in a hierarchical organization, often in demanding environments. Note how well each Option uses your specific job skills. Also enter as a PRO or CON how well each Option will utilize your experience in the military hierarchy with its clear lines of authority and communication.

3.1 G – CHANCE OF ADVANCEMENT OR LEARNING NEW SKILLS

Look at each of your Options in terms of advancement (which you've already tagged under Income) and in terms of the new skills you can learn (which will increase your ability to earn). From the official view of each Option, and from what you've been able to glean from talking to people, note probable training opportunities and promotions likely to follow such training.

> *Beware of all enterprises that require new clothes.*
> *— Henry David Thoreau*

For fields where the influence of politics or of powerful people is strong, note PROs and CONs. Especially be alert to Options where a network of former military people is available for learning the ropes.

3.1 H – ROOM FOR MAKING YOUR OWN DECISIONS

Look at each Option in terms of how active you could be in making decisions regarding (1) the goals of the work, and (2) the specific steps to be used in performing the work.

3.1 I – CHANCE TO FULFILL GOALS OUTSIDE OF WORK

Go through the Options and estimate how much leeway you will have in setting your own schedule. Try to predict how much you will be at the beck and call of production and deadline schedules without warning and how much these might interfere with what you like to do when you're not working. Also note which Options have work positions that overflow into the community (e.g. job-related sports teams, charity work, and social events).

3.2 – SELF-APPROVAL AND DISAPPROVAL

Some categories of anticipated outcomes are related to your attitudes about yourself. Here you look ahead for conditions that make you feel great and no-so-great about yourself.

3.2 A – HATING QUITTERS

You may have a strongly-held belief on when it is honorable to leave the military and when it is not. Perhaps you feel it is okay to leave only when physically disabled, after a given number of years of work, or when you reach a particular rate. Clarifying which conditions are right for you will help you feel good about whatever decision you make.

Part of the overall CON of leaving may be that you basically disapprove of people who leave anything. Especially if you have not served your full 20 years, you may be picturing yourself as a drifter, unloyal, or a job-hopper. In the civilian sector, there is an increased respect for people who change companies and jobs as part of an overall career-building plan.

3.2 B – MEANING OF THE WORK

The CONs against leaving the military are many, and changing from the status quo is likely to be an uphill battle because of the many predictable losses. A big CON can be the loss of clear values and mission, of knowing the ultimate purpose of why you go to work in the morning. The military certainly does not leave you wondering about the ultimate purpose of your work, and this may be one of the aspects of military life that is hardest to replace in the civilian sector.

3.2 C LINKS TO LONG-TERM GOALS

Many people have a clear picture of what they want to do after they leave the military. You might be willing to put up with a number of short-term difficulties that go with various Options because they lead, in the long run, to the final goal of your dreams. Yes, you may be willing to be the handyman at a resort for a while if it gets you the information, opportunity, and experience to purchase your own long-term dream of a fishing hole with a motel.

3.2 D – NEED TO COMPROMISE

Every job includes things you'd rather not do. This item is where you pay attention to those Options that ask you to compromise your basic values. Carefully note the PROs and CONs here, especially if you are an outspoken type.

3.2 E – CREATIVITY OF WORK

Sometimes, the degree of actual creativeness is matched by the amount of stress involved. There are many situations, especially when you include retirement Options, where opportunity to be productive or creative is a main attraction. Sometimes, however, high creativeness means high stress. As you enter your PROs and CONs for each Option here, be careful to readjust the Stress category (3.1 D) if it's changed by the creativity factor.

3.3 – PRACTICAL OUTCOMES FOR OTHERS/FAMILY

An important group of anticipated outcomes are the effects of your decision on other people. Because the effects on family members are likely to be intense, this section is written with them in mind. For your own future peace of mind, be sure to discuss these ideas with anyone who is involved in or will be affected by your decision.

Some of the practical effects on other people may be similar to the practical effects on you. But others in your family may have different opinions on whether you should stay in or not, and, if you get out, what they think you should do and where. Assuming that the approval or cooperation of family members and others is valuable to you, their specific wishes become PROs and CONs for each Option.

You may feel indebted to others who have put your career first up until now. You need to be aware that, especially in families, there is likely to be discomfort from feeling that it's payback time. Research shows that the payback may be in the form of restriction of freedom of action, loss of power and status relative to other family members, as well as actual dollar costs of supporting other family members' career needs.[24]

3.3 A – INCOME

One of the probable consequences of a family member leaving the military is that others in the family may have to go to work to bring in enough money for the family to make it. If another family member changes their life drastically depending on whether and when you leave the military, much discussion may be needed to work these plans out. If plans and feelings are not discussed up front, the partner may try to hide his or her success or failure at bringing in money.*

There are also likely to be many hyped-up positive feelings about Options that bring a big jump in pay. Consider these reactions in the context of the role money plays in your family as a whole.

3.3 B – STATUS

Does your family have a long-postponed dream that has been waiting until your military career is over? Have you been promising your spouse that "someday" the two of you will "have it made" and be free to live out this dream?

Enter items in terms that fit your family and the kind of status needs that they might want. Factor in that particular brand of status that goes with being in the military in any community, as well as how much of this is lost in each civilian Option.

3.3 C – TIME FOR FAMILY AND OTHERS

Wanting to have more family time is a common hope of military personnel who have been away from home a lot, and can be a PRO for civilian Options and some military Options. However, just because you now would like to spend more time with your family doesn't mean they are ready to spend more time with you. Children who have grown up without you around much will need to get to know you in a new way. Each of the parties will have to work at creating a new style of relationship.

If you will be around home a lot more, this will have an impact on your family. If you will not have a full-time job to occupy you, this impact will require significant adjustment by all. As one spouse put it, "...his arrival is felt as some sort of natural disaster." Family members tend to scatter when they feel they are expected to be a

* **"When He's Without a Job..."**: Reporting on executive retirement, Baker notes, "And another wife, who worked during her husband's bout of unemployment, hesitated to say she was tired at the end of a long day for fear of 'rubbing salt in his wound.' She'd get dressed in the bathroom each morning and practically sneak out of the house so he wouldn't feel bad that she had a place to go and he didn't." [25]

home version of a military staff group. Families can make the necessary adjustments, but only if they are ready to work through disagreements during the transition period. You'll all be better off if you don't expect these changes to just come naturally.

Adjusting to more time with the family comes easier if you have not overly used your military work as a shield to deflect involvement in family activities and problems. If your career included alternating times of high and low family involvement, the pattern for adjusting to your future role is already set. Your next career change can be adjusted to as just another return to high family involvement on your part. If you've been largely uninvolved, a new pattern for your participation in daily family life will have to be created.

3.3 D – DECISION-MAKING STRESS

If you have the kind of relationships where major family decisions are worked out in full partnership, decision making will naturally include your spouse and children. Remember, your spouse and family members may have put careers on hold as you were moved here and there with little warning. By the same token, if you have made most decisions for your family, this is a golden opportunity to readjust the way you make decisions, including spouse and family in the process. Even if you encourage their participation, however, they may hold back in order to avoid the responsibility if things don't turn out well. If you and your spouse share the decision, be prepared for some additional work. When two people make balance sheets for the same decision, you may initially have twice the confusion instead of twice the information.[26]

> The best career advice given to the young is 'Find out what you like doing best and get someone to pay you for doing it.'
> — Katherine Whitehorn

If you have made joint decisions before this, and you now revert to treating this decision about staying in or leaving the military as an individual one, your partner may feel very left out. If you have little experience in joint decision making, trying to be co-deciders for the first time on such a big decision may lead to major blowups that don't go anywhere. Watch out for compromises where no one gets what they want. Also, if you're heading toward having a wife make choices because up to now the choices were based on the husband's career needs, make sure you both agree on this arrangement.[27]

3.4 – OTHERS' APPROVAL AND DISAPPROVAL

This is the section where you try to predict how other people will feel about your decisions. The people involved may be different for each Option. For example, you will want to think about how your current co-workers would respond if you shift to another program in the military.

The family needs to be considered in two ways. Not just how each member might approve or disapprove of each Option, but how they might also approve or disapprove of how this decision is made. Generally families have their own history of how big decisions get made that will affect everyone. Various factors are included, even if not spoken about openly: who makes which decisions, any ground rules that limit them, how information is gathered, the types of appeals and safeguards provided for, and how the family actually moves into implementing change once a decision is made.

How do you tell how people are going to react? It's important to think this part through because we all worry about what someone is thinking about us. But remember, once we start predicting what people will think, do, or say, we tend to forget that we made these things up in the first place. We begin to believe our own predictions. Second-guessing other people or playing mind-reading games can keep you thoroughly stuck and miserable. Worse, you can do this all by yourself. Until you actually talk it out with the other person, you're at the mercy of your imagination.

Overpredicting the reactions of others will need to be handled as part of the process of leaving. Certainly you should try to predict the reactions of people that are important to you. But you will need to respond to their real reactions, not the ones you think they have.*

From your co-workers, bosses, and subordinates, you might predict varied reactions to your Option of leaving the military, from a large sigh of relief to feelings of sadness. But your prediction might degenerate into questions like, "What if they think that I...?" or "What if they say that I...?" These what-ifs can stall action and carry

---

* The Lawnmower Story: Jerry Trapasso was cleaning out his garage on a Saturday afternoon. He set his beer on the workbench to climb behind an upright mattress and see what had been abandoned back there. A posthole digger, a half-used bag of lawn feed, a broken bird house, and the lawnmower. Whoops. Not their lawnmower!
This was his neighbor's old push mower that he'd borrowed back during the gas crunch and obviously (now!) never returned. He sighed and began moving the mattress, untangling the other tools, thinking, "Well, how come he never reminded me anyway? You'd think if he'd needed it, he would have called or something. He probably hadn't needed it anyway...he probably couldn't remember who had it! Maybe he thought his neighbor on the other side had it, or the guy across the street. He really didn't keep very close track of his stuff. If he cared, he should have come and found it."
Jerry is pushing the mower, clacking up his neighbor's front walk. Rings the doorbell. "Maybe he's not home. Shouldn't lose track of his stuff like that!" Neighbor opens door sleepily. "Hey, Jerry. What you got there?" Jerry thumps it on the handle, "Here's you're *#@%/&%$#/@ lawnmower, Herb, and I sure hope you take better care of it next time."

the risk of your beginning to believe that such creative predictions are true. Here is where you pencil in these predictions, *knowing* they are only your predictions, not reality.

3.4 A – PRIOR COMMITMENTS

If you've been telling everyone you know that you're going to handle your decision in one particular way ("I'm staying in for the full twenty because it's obviously the smart thing to do."), and are about to change your mind, now is the time to face up to it. You need to be free to look at each Option for the costs and benefits it carries in its own right. If you've been mouthing off for years about how sure you are about what you're going to do, you're now going to claim the right to **change your mind**. You may need help with this. You may need support to face those who love seeing someone else have to admit they were wrong about something.

Depending on your partnerships, your marriage, your role as a parent or as a son or daughter, you have made a number of prior commitments to others, some spoken, some not.* Like the time I urged my parents to move three thousand miles so we'd be closer as they got older; my implicit commitment was that I wouldn't then pack up and move away, and that I had the power in my life to control such moves.

Up until now, you may not have made many commitments based on where you live simply because you never could count on where you'd be, and it was pretty much up to Uncle Sam to have the final say as to where you'd go. Now consider each Option in terms of how it will impact these spoken and unspoken commitments to the others in your life.

3.4 B – STATUS

There are others whose opinions of you matter to you simply as a matter of status. There are some situations you'd rather not be in just because the status that goes with the slot is just too low (or too

---

* **Spouse Reactions:** George Kennedy became an actor after being in the Army since he was seventeen, fourteen years of service. He explains that staying in had been easier than facing the unknowns of getting out. "We go along like a train. We have a tendency to stay on the same track unless something happens to disrupt that. And that's not always something from within ourselves. My wife used to say to me, 'What is this ACTOR? You can't be an actor. You can't do anything but what you're doing. Just stay in the service...' in effect until you die. She had no sympathy at all with the idea that I wanted to act. I think it was the lack of security more than anything. We had little enough the way we were so why take a chance?"
George Kennedy's spouse certainly disapproves of his Acting Option. Part of this stems from the implicit prior commitment she felt he made to support her securely.[28]

high) to be comfortable. Consider each of your Options in terms of the status associated with it, not only after you've implemented the Option, but also throughout any steps you'll have to go through to get there.

Over the last ten years, there have been changes in the American culture that generally make it easier for people at mid-life to retrain for new careers. College campuses welcome returning students, and vocational and career counselors are getting more and more experienced in helping people make mid-life career changes. In addition, most adult college students are also working at a salaried job, so that you are less likely to end up sticking out like a sore thumb in a school setting. This increase in sequential careers for all kinds of people changes how ex-military folk are received in the workforce. Certainly, you won't stand out so much as in years past.

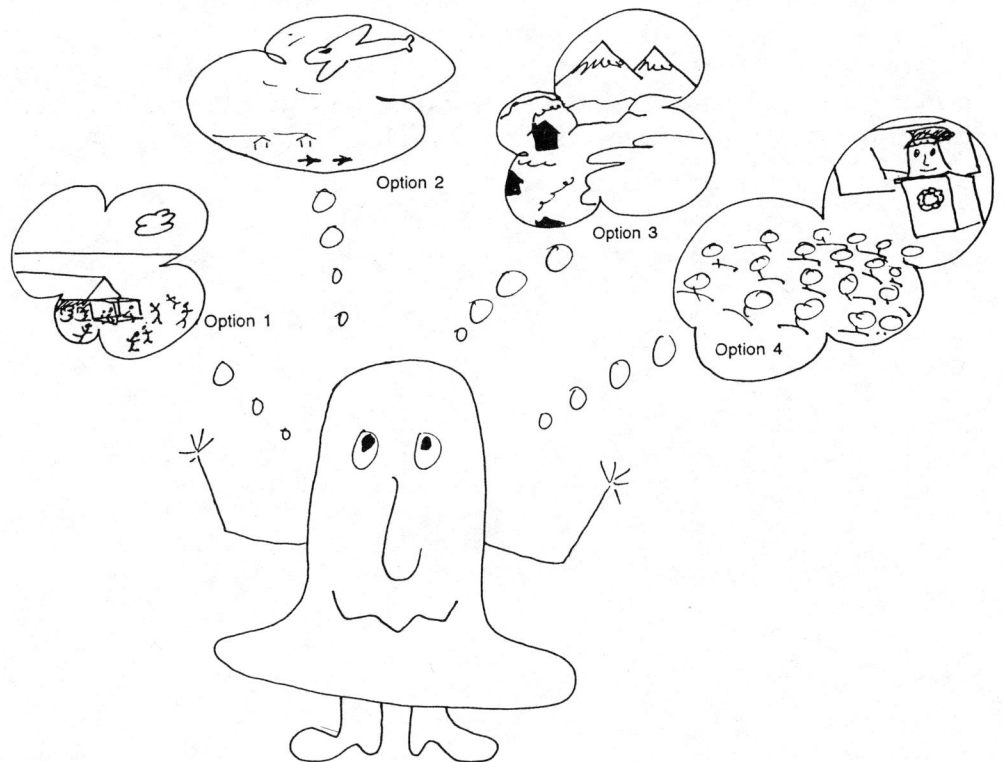

Figure 3–1: Balancing Multiple PROs and CONs

The loss of status of being in the military may be important. You have to wonder what it would feel like not to have that pay rate, job, and position as a part of who you are. If you have had a good deal of power, prestige, and authority, they have become part of your self-image. Even if your status has not been high, you may miss the clarity of status in the military world. You may have to adjust to the vagueness of civilian status.

3.4 C – INDEPENDENCE VERSUS RECOGNITION

Look at your Options in terms of how much approval or disapproval of others is likely. One of the CONs for re-entry into the civilian sector is a continuing prejudice that ex-military folk have fat retirement checks coming in at taxpayers' expense. A guide to facing the PROs and CONs of a military background is provided in *Re-Entry: Turning Military Experience into Civilian Success* by Keith O. Nyman (see Appendix A).

You need not take stock here of how important the recognition of others is to you; you will get to weigh these various kinds of approval of others in Chapter 4, where you balance out your personal preferences. For now, you need only enter the PROs for each Option that provide recognition by others or freedom to work on your own.

CHAPTER 4

STEP 4 — SUMMING UP

Look at your Trial Balance Sheet. How are you going to pull all this together? This chapter provides specific directions for creating a Final Balance Sheet of your Top Three Options.

To prepare for summing up, reproduce a few copies of your Trial Balance Sheet so you can mark them up while going through this chapter.

4.1 – WEIGHTING YOUR TRIAL BALANCE SHEET

To see The Big Picture in perspective, pull together all the information you have gathered, following the simple steps provided in this chapter.

Even if you feel you have already decided what to do, it will be valuable to apply these next steps to carry forward what you have learned into the implementation and transition phases of your decision.

☆ ☆ **WORK WITH TRIAL BALANCE SHEET** ☆ ☆

4.1 A – MARK NON-NEGOTIABLE ITEMS

1. Check each of the four sections of the Trial Balance Sheet for non-negotiable outcomes.

2. Are there any PROs that you can't live without? Draw a box around them in red.

3. Are there any CONs that you can't live with? Draw a box around them in red.

4.1 B – CHOOSE THE EIGHT MOST IMPORTANT ITEMS ✎

1. Scan your entries in the first section of the Trial Balance Sheet — Practical Outcomes for You.

2. Circle in red the PRO that is the *most important to you*. Ask yourself: If you could only have one PRO, which would you take?

3. Circle in red the CON that is the *most important to you*. Ask yourself: If you could only avoid one CON, which would it be?

4. Scan your entries in the second section of the Trial Balance Sheet — Self Approval and Disapproval.

5. Circle in red the PRO that is the *most important to you*.

6. Circle in red the CON that is the *most important to you*.

7. Scan your entries in the third section of the Trial Balance Sheet — Practical Outcomes for Others/Family.

8. Circle in red the PRO that is the *most important to you*.

9. Circle in red the CON that is the *most important to you*.

10. Scan your entries in the fourth section of the Trial Balance Sheet — Others' Approval and Disapproval.

11. Circle in red the PRO that is the *most important to you*.

12. Circle in red the CON that is the *most important to you*.

4.1 C – PUT THE EIGHT MOST IMPORTANT IN ORDER

On your Trial Balance Sheet, you have circled eight possible outcomes that are important to you. Next, take a short break to shift your focus as you prepare to see the Big Picture. Stand up, stretch, look out a window, get a drink of water.

Now, scan the eight items, looking only at them, and mark a number beside each in order of importance to you. For some people it is easier to start with the number-one outcome; for others it's easier to start with the least important of these eight and work upward.

If you need help, you can use a comparison chart. There are several logical ways to compare each pair of Outcomes to come up with a single series numbered in order of importance to you. Bolles shows one way in his book *The Three Boxes of Life*;[29] another is in the following FYI box.*

---

* Comparison of the Eight Most Important PROs and CONs

	A. No sports money – CON	B. With friends – PRO	C. No soccer for son – CON	D. Medical benefits – CON	E. Lifelong dream – PRO	F. Politics on the job – CON	G. Wife's short commute – PRO	H. Wife's long commute – CON
A. No sports money – CON								
B. With friends – PRO	B							
C. No soccer for son – CON	A	B						
D. Medical benefits – CON	D	D	D					
E. Lifelong dream – PRO	E	E	E	E				
F. Politics on the job – CON	A	B	C	D	E			
G. Wife's short commute – PRO	A	B	C	D	E	F		
H. Wife's long commute – CON	A	B	C	D	E	F	G	

Starting at the top left, compare A to B, and decide which is more important to you. Write its letter in the comparison box where the row and column intersect. Here, a person compares the CON of no money for sports (A) to the PRO of being with friends (B) and chooses B as more important. Which of two PROs would you rather have? Which of two CONs would you rather do without? Or, when comparing a PRO and CON, would you rather give up a PRO or get out of a CON? When comparing two items that have to do with Others' Practical Needs or Approval and Disapproval, decide which is more important to the others involved. If there are different Other people involved in the two outcomes being compared, decide, in a general way, which person is more important to you. Sometimes two items are the flip of one another; in the example above, you can see that G and H are very similar. Although one is a PRO and one a CON, the sense of each is similar, and they have the same relative weighting against the other outcomes in terms of importance. This person, when forced to choose one, picked G to enter into the comparison box.

When you've finished your paired comparisons, tally up the total number of times each Outcome was chosen. Count the number of A's and note it, then the number of B's, etc. through H. In the above example, the totals are A=4, B=5, C=3, D=6, E=7, F=2, G=1, H=0. From this tally, you can now number the Outcomes from the Most to Least Important, with the highest number showing the Outcome with the most votes. In this example, E is Most Important, then D, then B, then A, then C, then F, then G, and Least Important is H.

4.1 D – IDENTIFY THE TOP THREE ITEMS

However you achieved it, you now have an idea of which Outcome items weigh the most with you. Go back with a highlighter and highlight the top three items. Next, search all your other entries to see if these same items appear in any other spots on your Trial Balance Sheet and highlight those as well.

☆ ☆ REVIEW TRIAL BALANCE SHEET ☆ ☆

Now fill in the following three sentences using the top three of your eight most important items (an extra form for this exercise is provided in Appendix C):

The *first* most important thing to me is to

      get _____ (+3 points)

or to avoid _____ (−3 points)

        an issue which is practical _____

     or one of approval/disapproval _____

                   for me _____

              or others _____

The *second* most important thing to me is to

      get _____ (+2 points)

or to avoid _____ (−2 points)

        an issue which is practical _____

     or one of approval/disapproval _____

                   for me _____

              or others _____

The *third* most important thing to me is to

      get _____ (+1 point)

or to avoid _____ (−1 point)

        an issue which is practical _____

     or one of approval/disapproval _____

                   for me _____

              or others _____

This exercise exposes two major aspects of what's most important to you: (1) whether they are PROs or CONs, and (2) whether they have to do with you yourself (sections A and B on the Trial Balance Sheet) or with other people (sections C and D on the Trial Balance Sheet). Find the Options for these Top Three Outcomes on your Trial Balance Sheet and give three points for the most important, two for the next, and one point for the third.

4.1 E – CATCH YOUR REACTIONS TO THE TOP THREE

By now, it may be clear which Option seems to be coming out the winner on paper! Add up the values of the Most Important PROs for each Option as a plus score; add in the points for the top three items you noted in the last paragraph (4.1D). Then add up the Most Important CONs as a minus score, and mark the plus and minus scores at the bottom under each Option. If the plus and minus scores clearly point to an Option that feels right to you, mark it Option Number One.

If the scores indicate a tie between two contending Options, or, if you feel more attracted to the second-place Option, look at the scores more closely. Notice if one contender is highly polarized (with high plusses and minuses) and the other is less polarized (e.g., lower or more equal plusses and minuses). Mark the more polarized Option "More Risky" and mark the less polarized Option "Less Risky." Now see if you can select between the contending Options with more confidence.

If you are still uncertain, flip a coin to decide between two contending Options. Let the coin tell you which Option will be Number One. In the next section, there will be more tools for breaking these logjams.

After noticing your reactions to any Options that are close, now notice your reactions to the various boxes, circles, highlightings, and numbers. Pay attention to the thoughts that come to you! Write them down. If your thoughts and emotional reactions are going in the same direction as the circles, underlinings, and highlightings, you're close to a decision. If your thoughts and feelings are in some other direction, note which Option you wish had turned up the winner on paper.

If you're pretty sure which Option you're going for at this point, you can jump ahead to Chapter 6 – Carrying Out Your Decision, to transform your current thinking into useful tools for the implementation and transition phases of your decision. If your circles and lists are still confusing, the next sections provide more assistance.

4.2 – REFRAMING YOUR OPTIONS

4.2 A – NON-NEGOTIABLE ITEMS

Look at each non-negotiable item. If there are PROs that you are unwilling to do without, can you think of any way to add the same or similar PROs to other Options? Is there any way to change the other Options you're considering to include more of what is important to you? For example, can time off for a son's wedding be attained with any other Option by any means whatsoever? Then, look at each CON you won't negotiate on, won't put up with. Look at where they appear, and think about that particular Option. Try to come up with a way to change the Option so you can avoid the CON or change how you'll handle it to reduce its painfulness.

4.2 B – WORDING

Now work on the wording of each Option so that it is short and clear. Especially change the wording about timeframes until you have three or four Options that are crystal clear about "what" and "when."

4.2 C – RECOPY OPTIONS

Start over with a clean sheet labeled Final Balance Sheet (reproduce the copy provided in Appendix C). Start with your Options and re-write them in their best wording, with your first choice in the left-hand box and the other Options to the right, in order of choice. Do this even if you are working with a First Choice Option picked by the flip of a coin.

☆ ☆ COPY ONTO FINAL BALANCE SHEET ☆ ☆

4.2 D – RECOPY ANTICIPATED OUTCOMES, PRO AND CON

Once your Options are fine-tuned, go ahead and redo your PROs and CONs from your Trial Balance Sheet but fine-tune these also, as you go.

4.3 – DEFENSE MANEUVERS

If your work on paper is not clear, note where the conflicts are. Do you come up with one Option for your own needs and another for other people in your life? Do you come up with one Option in terms of practical needs and another in terms of personal approval and disapproval? Try to characterize the conflicting nature of these Options in one brief statement. For example, "Financially, I choose Option 1, but Option 2 would be more personally satisfying." Or, "I don't know whether to put up with the hassles of Option 1 or to go for the extras of Option 3." Once you've focused this final choice point, you can use the two tools presented in this chapter to break the logjam. Using Counterweighting to Correct for Personal Style, you will learn to understand your own personal style of explaining outcomes and how to borrow the strengths of other styles to maximize your PROs and minimize your CONs. Using External Calibration, you will discover ways of using input from other people.

4.3 A – COUNTERWEIGHTING TO CORRECT FOR PERSONAL STYLE

Locus of Control – Locus of Control describes how different people explain to themselves the outcomes in their lives. For your purposes, it will be helpful to know how you explain the outcomes of your past decisions and the one you're working on now.

Researchers[30] found that people attribute control of the outcome of events in their lives to one of three basic "locations": Internal, External/Chance, or External/Powerful Others.

1. People with an **Internal** locus of control attribute outcomes to their own actions. "I won the bowling tournament because I really put in the practice time."

2. People with an **External/Chance** locus of control attribute outcomes to outside influences or to chance. "If the weather hadn't changed, I never would have won." (External) "If it hadn't been such a bad week, I would have won." (Chance)

3. Finally, people with an **External/Powerful Others** locus of control attribute what happens in their lives to the actions of other people. "That coaching session with Ed really gave me the edge I need to win."

People *believe* that outcomes of events are controlled in this way. As we will see later in Chapter 5, any belief has the potential to become a trap unless you understand the role it plays in your life. Once you understand the pitfalls that go with each of these beliefs, you can take steps to counteract them. Before dealing with these pitfalls, take a minute to decide your style.

Deciding your style – If you believe that control of what happens in your life is mostly a function of what you do, you are an Internal. If you see events in your life as controlled from outside yourself, you are an External. External control can be seen as due to factors such as chance or unknowable events, or to powerful others such as authorities and experts.

Think about three people who believe in God but have different ideas about how God operates. The believer with the Internal style might say, "God helps those that help themselves." The believer with the External/Chance style might say, "God works in mysterious ways." The believer with the External/Powerful Others style might say, "God will provide." You can see how these three orientations would create different approaches to living and to personal decision making.

You can use the information generated in your Lifemap and your Trial Balance Sheet to discover your Locus of Control style. You may notice a change in style over the years or at various times of your life. Especially notice times where you felt that the outcome didn't have much to do with your input. If you need more help in determining your general style, take the Locus of Control questionnaire in Appendix B.

Unlike laboratory experiments, in life you *rarely* have **nothing** to do with the outcome. Choosing not to choose is in itself a choice of the status quo. It's especially important to notice to what you attribute your past failures and successes. Some people have an interesting pattern of attributing all their success to luck or chance and all their failures to their own actions. You need to know if this is your pattern. By the same token, the pattern of attributing all your success to your own actions and any failures to powerful others is just as one-sided! (More on this under Double Standards in Chapter 6.)

As you think about your style, you may realize that you are one type for one area of your life and a different type for another. For instance, a college student might be Internally oriented around looking for a summer job, "If I look long enough, I'll find something," but externally oriented in terms of grades in class, "That teacher is really tough!"

Decide what type you are or take the test in Appendix B.

☆ USE LOCUS OF CONTROL SCALE (IF NEEDED) ☆

Characteristics of three Locus of Control styles – We've already discussed how a feeling of control over outcomes can reduce stress. This book provides tools to increase control over the decision-making process. Using tools from a book would be more natural for those who have an Internal locus of control orientation. They already feel they can influence the outcomes by what they do. If you've read the book this far and carried out the exercises, you are approaching this decision-making task in an internally oriented way, regardless of how you act in other areas of your life. It seems likely that Internals are the only ones that follow psychologists' decision-making advice since "...all the refined decision-making techniques ever developed would do little to aid the individual who believes that outcomes in life are largely beyond personal control." [31]

In general, your style (Internal, External/Chance, or External/Powerful Others) will be seen as positive or negative depending on the setting in which you are operating. How well your beliefs about control match up to the external situation will affect your stress in decision making. Research implies that Internal people are somehow "better," but life is not that simple. In fact, most of this research comes from universities where researchers "publish or perish" based on their own output. Such an atmosphere, where there is little organizational regulation, promotes an Internal orientation, and the research favors this Locus of Control. By the same token, the more structured military organization promotes an External/Powerful Others orientation. Near the end of this chapter (Matching Personal Styles to Options), you will find ways for checking out each of your Options in terms of how well they match up with each Locus of Control style.

Because Internals assume that outcomes are a function of their own behavior, they are likely to do everything they can to maximize the PROs and avoid CONs. An Internal might even selectively notice instances when their actions affect outcomes and selectively ignore instances when their actions don't affect outcomes. Internals are likely to overcontrol. For example, an Internal is likely to keep driving to more banks to find one open on a given Monday, while an External would more quickly look for reasons in the environment, more quickly realizing that it's a holiday.

Some people attribute outcomes to chance. As Friedrich put it, "In the extreme, the **external** presumably is not concerned with taking actions designed to control his reinforcement; he may be relatively insensitive to whether or not a decisional situation allows a sense of freedom. In a simple situation like choosing white meat or dark meat, an external would be more likely to take whatever is easier for the server." [32]

> *The Serenity Prayer…*
> *God grant me*
> *the Serenity to accept*
> *the things I cannot change,*
> *Courage to change*
> *the things I can,*
> *and Wisdom to know*
> *the difference.*
> *—Anon.*

> *Circumstances are the*
> *rulers of the weak;*
> *they are but the*
> *instruments of the wise.*
> *—Samuel Love*

There are other ways to view Externals who see chance as powerful in determining outcomes. Such an emphasis may be a mark of a person feeling helpless or a person who is sensitive to the timing of life's events.*

Among people who attribute outcomes to external factors, there are those who focus not so much on chance or random factors but on the influence of others, those assumed to be powerful. Because of the categories used in your Trial Balance Sheet, be aware of this distinction. The power of others can relate to skills, access to resources, or power in relationships. An example of relationship power is the child whose special medical needs require the family to be near a metropolitan hospital; here the child is a Powerful Other because the family cares about her and her needs.

ESCAPE HATCHES FROM YOUR OWN PERSONAL STYLE

Knowing your type allows you to steal the perspectives of the other two types to balance out your weak spots. If you don't take time to walk around your Final Balance Sheet and look at it from several perspectives, problems hidden by your blind spots will come up later and throw you.

One way this review can help you fine-tune your Final Balance Sheet is in reframing your Options. Bring yourself back to a close choice, a choice between your top two Options. Now re-read a description of one of the Locus of Control styles that is **not** yours and imagine which Option that person would choose. Do this again with the second Locus of Control style that is **not** yours, adding new ideas to your Options and Final Balance Sheet entries to counterweight your normal Locus of Control style.

---

* **External Sensitivity to Timing as Described by Winnie the Pooh:** "...you go by circumstances and listen to your own intuition. 'This isn't the best time to do this. I'd better go *that* way.' Like that. When you do that sort of thing, people may say you have a Sixth Sense or something. All it really is, though, is being Sensitive to Circumstances. That's just natural. It's only strange when you *don't* listen.

"One of the most convenient things about this Sensitivity to Circumstances is that you don't have to make so many difficult decisions. Instead, you can let them make themselves. For example, in *The House at Pooh Corner*, Pooh was wandering around one day trying to decide whom he wanted to visit. He could go see Eeyore, whom he hadn't seen since yesterday, or Owl, whom he hadn't seen since the day before yesterday, or Kanga, Roo, and Tigger, all of whom he hadn't seen for quite a while. How did he decide? He sat down on a rock in the middle of the stream and sang a song.

"Then he got up and wandered around again, thinking about visiting Rabbit, until he found himself at his own front door. He went inside, got something to eat, and then went out to see Piglet.

"That's how it is when you use the Pooh Way. Nothing to it. No stress, no mess." [33]

Retune these one more time and imagine the impact of each worst outcome. Then put in the steps you would have to take in order not to get caught any shorter than necessary if the worst occurs. In the same way, if there are steps you need to take to be able to utilize the best possible outcomes (e.g. filing for an independent business license just in case), write them into your Option statements.

There are several other ways to counterweight for your natural decision-making style:

- Focus on predicting the "worst" possible outcome as a CON, and the "best" possible outcome as PRO for your top two Options.

- Mark PRO or CON entries that you feel are influenced by your Locus of Control style. Circle them and then look over the whole Final Balance Sheet to see if they cluster in any particular area. Are they mainly personal concerns (self) or to do with others? Are they practical or are they mainly concerned with approval and disapproval?

Knowing your own Locus of Control style, notice specifically the pitfall areas* that go with it and build in at least two ways to minimize these particular pitfalls. In addition, look at the plusses of the other two styles and create at least two ways to get more of those styles' plusses into your decision making.

MATCHING PERSONAL STYLE TO OPTIONS

Back in earlier chapters, different types of control were discussed as we looked at the Control Dam in the Stress Landscape. Three main ways of minimizing stress were discussed: being motivated and committed to a goal, being able to predict events or outcomes, or being able to reverse them.

To apply these ideas to your current decision of whether to stay in or leave the military, rate (on a scale of 1 to 10) how much control you feel your actions will have on the specific Options you are considering. If one of your Options is to go into a business that requires a computer, for example, the degree of control you now have is greater than it would have been 10 years ago when only people who

* Plusses and Pitfalls of Three Locus of Control Styles		
	Plusses	**Pitfalls**
Internal	High motivation	Doesn't factor in Others' influences; Slights contingency planning
External/Chance	Sensitive to timing	Gets discouraged; Doesn't factor in Others' influences
External/Powerful	Reads Powerful Others well	Gets discouraged; Slights contingency planning

were pretty wealthy could get their hands on both computers and the training required to use one. Similarly, if one of your Options is to become some kind of professional sports person, your own skills and background will tell you how much control you're going to have over that outcome, based on the skills and experiences you already have.

Finally, you can do a mental experiment by comparing the style of your particular job in the military against the Options you are considering in terms of which is easier* for Internals and which for External/Chance and External/Powerful Others.

4.3 B – EXTERNAL CALIBRATION

SOUNDING BOARDS

By now, you may be even less clear among your Options because you've been focusing on your own personal style and may need to step back to get a larger perspective that includes a wide range of views. Here are several ways to get others to help you do it.

Sounding boards** are people we ask to listen to our thoughts, not to tell us what to do but to help us find the flaws and gaps in our own thinking. We are not asking them for their opinions. You have to be free enough with them that you won't feel frozen into any decision just because you were excited about it when you talked it over with them. A sounding board has to help you feel free enough to change your mind.

---

* The Effect of the Environment on Three Styles: In one study, comparing only Internals and Externals, the researcher thought that Externals would show less stress in departments of a hospital that were more formalized and structured while Internals would show less stress in the more fluid, unpredictable departments with more varied work.[34] This was found to be true only when departments varied in job specificity, that is, having specific procedures for any situation and putting stress on going through proper channels. He didn't find his idea held up because of other aspects of formal organization.

** Mt. St. Helens: To play Mt. St. Helens, you need to find a patient friend who is willing to listen to you for five minutes without interrupting. Your job is to erupt. The listener is not to interfere with your erupting!

Set a timer and suggest that the listener have paper and pencil to write down their questions and comments. This will help them remember questions to ask you later to help you formulate your ideas more clearly once the eruption is over and it will give them something to do with those persistent comments that are on the tip of their tongue. It'll help them keep from interrupting you.

Start the timer and begin to go through your whole set of ideas about the PROs and CONs of your decision or any part of the process that has been giving you trouble. When your time is up, sit back and let your sounding board either just hand you their questions so you can go over them by yourself later, or you may want to go through them one by one, seeing how clearly you can state your thoughts. Make notes yourself of any gaps in missing information pointed out by your sounding board so you can fill them in later.

Their role is to get you to state your thoughts more clearly, not attack your ideas. As soon as their questions feel like an attack in the slightest way, you're pushed into defending your position. People have been pushed into prematurely making a decision just because they found themselves defending some of their momentary enthusiasms that were under attack. A sounding board needs to avoid pushing you into solidifying your decisions, whether by asking attacking questions or by showing their impatience.

THE DEVIL'S ADVOCATE

You probably have several friends who love to debate and argue, who have a wide streak of lawyer in them, who will be glad to play Devil's Advocate once you want to try standing up for a decision. You will learn a lot by doing this for your top two Options even if you have already decided on your Chosen Option.

Have the Devil's Advocate attempt to poke holes in the logic you use to present your choice. Also, have him or her try to catch you in spots where you are not so convincing, based on your voice tone, your expression, or just a general impression that you're pulling your punches. Take notes on the points that raise doubts for you. If you tend to be an optimistic Internal, this process can alert you to many problems you wouldn't have thought of otherwise.

4.4 – THE FINAL DECISION

It is possible that after going through all the information on your Final Balance Sheet and fine-tuning with Locus of Control style and sounding boards, you still have not come up with a Numero Uno Option. There are several approaches to this dilemma:

1. Go through the four steps once more. Recopy your notes so they are clearer and cleaner this time. You may find that your Numero Uno pops out at you or becomes clear only slowly as you go through the steps a second time.

   a. Review your Lifemap for former decisions where you got stuck like you are now. Review Chapter 1.

   b. Recopy a new Table of Options. Review Chapter 2.

   c. Redo your Trial Balance Sheet. Review Chapter 3.

   d. Recalculate and reweigh for a Final Balance Sheet. Review Step 4.

2. Put your notes away for a while, let the tension build and watch your dreams.

3. Proceed to Chapter 5 to learn about Decision Traps and Escape Hatches and about problem-solving techniques to get back on track.

4. If all else fails, continue with both your first and second Options as you move into the implementation phase (Chapter 6). With this approach, whatever comes out as Numero Uno is your "plan," and the runner-up is your backup.

Whichever way you work next, it is important to stop reading at this point and label your top Option as Numero Uno, especially if you are going on to the tracking process of Chapter 6.

CHAPTER 5

DECISION TRAPS
AND ESCAPE HATCHES

This chapter deals with some of the problems you might have encountered as you worked your way through this book. Decision stresses, beliefs that can trap you, avoidance patterns, and timing problems are discussed in detail. It will be helpful to have your Lifemap to mark up as you read this chapter and identify your own traps.

5.1 – DECISION STRESS

5.1 A – INDECISION VERSUS KEEPING YOUR OPTIONS OPEN

Indecision is a different kind of stressor than those based on events that have happened in your life. When you can't decide between several Options, you have to face the stressors that come with each. Each Option is accompanied by possible outcomes that you don't want. Considering all the possible outcomes while you're in the state of indecision is often mentally more difficult than living with the worst outcome of any Option. At least you don't have to keep reviewing all the possible negative outcomes of all your possible Options.

On the other hand, being in the state of indecision also allows you to dream of all the possible wonderful futures you might have. Some people value this phase of their lives.*

Figure 5–1 is a flow chart of the different ways people can respond to indecision. It diagrams the decision-making process from the point when you first become aware of your dissatisfaction with the military life to the point when you decide whether to dig in and make a stronger military career or to turn in your papers.

First, there is the type of person who actively complains. Figure 5–1 shows that such complaining may be the mark of the "well-adjusted chronic complainer." Such folk, and we've all been there, provide a potentially useful escape valve for many work groups. These people live their lives paddling around in the eddies of indecision, and appear to be pretty comfortable there.

If chronic complaining does not serve the function of letting off steam, we must determine whether this person with complaints has faced the decision to leave or not. If they haven't seriously considered this obvious Option, they may be stuck in indecision without knowing it.

Once the decision to leave has been faced, some folks decide to leave but then wait for a variety of reasons. They adopt a short-timer's attitude, even if it goes on for years. People who have not decided to leave or stay need to evaluate their whole military career, explore new Options, create a new career plan, and decide whether to go for it or not.

If you are not actively working toward your decision, have made no decision to either stay or go, and avoid the whole subject, then you are likely to fall in one of three styles of being stuck: stuck and immobilized, stuck and swinging back and forth, or stuck and faking it by pretending that there's nothing going on. Each style of being stuck is likely to affect your performance at work, above and beyond those work problems that got you thinking about leaving in the first place.

---

* **Backwater Potential:** "We discern that there are, in general, two kinds of roads in our life; the roads we have actually traveled and the roads we did not take. What we found on the roads that we traveled is now known to us. Those untaken roads contain the experiences of our life that have remained unlived. In one sense they are now beyond us, like river water that has flown out to sea. But in a deeper sense they still contain many possibilities of life that are still present and available to us; it is these that we must explore" [35]

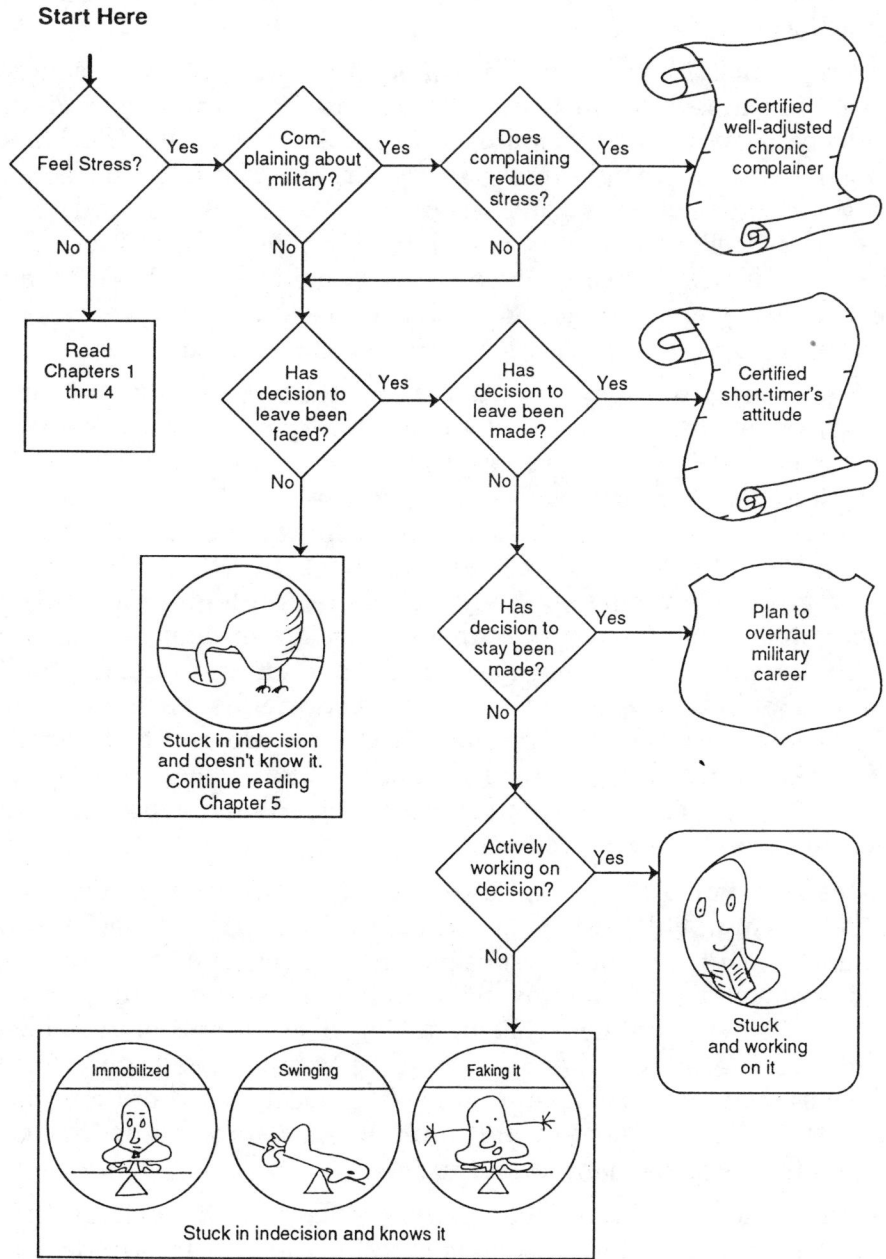

Figure 5–1. Flowchart of Indecision

Being stuck and immobilized magnifies any uncertainties about job performance in two ways. It changes your commitment to the job, causing you to feel that you may not be performing as well, and it undermines your self-image, causing you to misjudge feedback on your performance.

Being stuck in indecision is likely to lower job performance because of reduced commitment. Programs where you're not giving your all can get fouled up more easily and lead to the "I'll-leave-when-this-mess-is-straightened-out" syndrome. Finding yourself in the stuck mode sends out ripples of hesitation and self-consciousness. These ripples result in a considerable blurring of your image, if not your effectiveness. A growing lack of clarity and focus because you're tiptoeing around or because your moves are uncertain can quickly erode a formerly decisive stance.

When your attention is focused on deciding to leave or not, it is also harder to catch events that indicate people are losing trust in your judgment. When you keep getting distracted by thoughts of the future, feedback on job performance can get restricted to a very narrow time window. You may begin to forget important past or future factors because a voice in the back of your mind keeps reminding you, "You may not be here!" There is a high probability of overselecting either positive feedback ("I'm doing okay, things are better, I should stay.") or negative feedback ("I'm really doing badly so I might as well go."). At a time when accurate feedback on performance is most critical, the pressures to distort are greatest because you are constantly looking for information that leads to a "go" or "stay" decision. Then, without accurate feedback, you lose any chance to correct your performance.

It can be worse if, instead of being stuck and immobilized, you keep seesawing. This swinging back and forth between going and staying can result in wild fluctuations in your levels of effectiveness. For example, pessimism may be peaking on the day of a major meeting that was planned earlier during a more optimistic phase. You may have assumed that you would be able to speak quite easily off the cuff. Poor performance in a meeting because one did not allow for cycles of discouragement and dips in confidence add further to feelings of self-consciousness and failure.

Loss of self-confidence can only escalate with each of these cycles. A typical reaction to compensate for the effects of indecision on performance is to try harder, acting in a way to conceal any lapses from others. A feeling of "faking it" may increase from a "business-as-usual" level to a feeling of putting on a "close-to-panic" act.

> *Self-confidence is the first requisite to great undertakings.*
> — *Samuel Johnson*

> *When a firm decisive spirit is recognized, it is curious to see how the space clears around a man and leaves him room and freedom.*
> — *John Foster*

5.1 B – DECIDING ON YOUR OWN

The military is not the best training ground for making your own personal decisions. It's likely that the military devoted a lot of effort to train you in many work and management skills, but spent hardly any effort on personal decision making.

If the service had wanted you to have a manual on how to decide when it's time to go, they would have issued you one. Right? Not right! The military may be providing you with a lot of services and training but they are unlikely to have provided training in assessing the effects of the military on you. All those evaluations you've been through were aimed at assessing your job performance — how well you were serving the military. Now it's time to ask — how well is the military serving you?

The more you are trained in the authority and responsibility aspects of decision making within and for an organization, the less time you've been practicing deciding for yourself. Solitary decision making may feel so uncomfortable that you start thinking about something else, anything else that's distracting.

> *Doubt what you will,*
> *but never yourself.*
> *—Bovee*

Review how much experience you've had with solo decision making since you've been in the military. Remember, one way to reduce stress is by increasing predictability and control and, in this case, it helps to make decisions yourself since it gives you control over the directions you take and the timeframes you set. To keep solo decision making from adding to your stress, you want to retain the decision power but get as much help as you can along the way.

Go back and review your Lifemap entries for those decisions you made yourself (↑). Notice how many you made totally on your own and how many were made by you but with help and support from other people. Note who these people were and see if you can recognize what, in particular, they did that was helpful to you.

☆ ☆ REVIEW YOUR LIFEMAP ☆ ☆

The stress of decision making on your own can be increased or reduced by the roles other people play while you're in the middle of the process. These people may influence and be influenced by the process of deciding as well as by the final decision. Their involvement in the process can affect your stress.

If you've been working on this decision for a while, usually you've let some people know you are considering staying in or returning to civilian life. Many have advice, platitudes and, naturally, opinions. Not taking their advice and opinions can create friction, and such friction can whittle away at the willingness of family and friends to support you while you are stuck in indecision.

There is a difference between using family and friends for a sounding board about the job itself and using them as a sounding board for the decision to leave a job. People react to talk of leaving a job, especially one that carries lifestyle with it, as a threat to your relationship with them. If you leave, it may mean their leaving with you (most likely a family reaction) or their being left behind (a concern of friends). If you stay, they come to distrust your talk about leaving, complaining that it's boring or irritating. When a friend begins to challenge you to leave, you might suspect that they are tired of hearing what they now perceive as idle threats.

Family members may have the least tolerance for discussions about wage earners leaving their jobs. Teenagers may leave the room, and spouses may tighten a jaw muscle and change the subject. Frequent mention of how important continuity is for a high school senior, finally finding the right day care center, scout troop, annual summer plans, etc. should tip you off to a campaign to maintain the status quo.

Once your family and friends begin to complain about your indecision or to campaign for their own wishes, your support for taking time to think through your decision thins out rapidly. Family and friends may now become a source of stress and your natural reaction would be to withdraw from them.

If you began the four steps and are not yet finished or can't come up with your best Option, look back and see if you got derailed by some kind of decision stress. It could be either the strain of indecision itself, problems getting the help you need, or pressure from family or friends.

To reduce indecision stress it may be useful to:

- Take breaks from thinking about your career at all
- Set a deadline for a final decision
- Reduce stress in other areas of your life (review Chapter 1).

To improve the help you get from others:

- Decide who will and won't be a useful sounding board or Devil's Advocate for you (review Chapter 4)
- When tempted to talk to others about your decision, write those thoughts down instead
- If external factors are forcing you to make a decision under high stress, hire a vocational counselor to help you.

5.2 – BELIEFS AS TRAPS

These next three sections are to help with three types of decision traps that can keep you stuck with no decision. If you're having trouble moving through the four steps, you may be in one of these three types of traps: beliefs, avoiding, or timing.

5.2 A – THE "RIGHT DECISION" TRAP

The biggest decision-making trap of all is believing that there is only one right decision possible, so if you choose any other one you will (1) be wrong, (2) be sorry, and (3) be regretting it for the rest of your life.

Decision Trap Belief #1: In any decision, there is only one right choice; all the other Options are wrong.

Escape Hatch #1: Go for the *best,* not the right choice.

Decision theorists grade a decision by focusing on *how* the decision is made. They consider the best decision to be one based on the best information available at the time you decide. Monday morning quarterbacking doesn't count.*

> *Do I contradict myself?*
> *Very well, I contradict*
> *myself; (I am large,*
> *I contain multitudes).*
> *— Walt Whitman*

If you believe the world is a place where human situations can be black and white, you are more likely to be hoping for one right alternative among all the wrong ones. The way you believe the world is, in general, will affect how you view this specific decision-making task. If you hold strong, fixed beliefs about how the world operates,

---

* If-Only Stories: Every family has one, a story about a great land deal, mineral rights you could have gotten for a song, hula hoops, pet rocks, or other inventions you should have invested in. Uncle Harry or someone like him tells the story on family holidays how he could have gotten in on the ground floor on some business deal and how, if he had, he'd be sitting pretty now, boy-oh-boy. New kids coming along are a great excuse for families to retell these stories.

Are we teaching the next generation that if things don't turn out as we predicted, then we made a wrong decision? Just because what looked like a fluke turned out to be a money-maker, do we tell ourselves we screwed up, that our decision was wrong?

Actually, any decision, any alternative, has its good and bad consequences. Uncle Harry rarely tells the other side of what he did do with the money that he didn't invest, or what might have happened if he hadn't had the money. What if the payments had drained them for years? What if? What if?

The biggest psychological trap in decision making is believing there is only one right answer and all the other alternatives are wrong. When people focus on the negative parts of the choices they make, they'll end up feeling they did the wrong thing no matter how thoughtfully they decided.

The next time the family stories are going around about the cool million that someone could have made, maybe some grown-up will wink at the kids who are listening and say, "Wellllll, maybe...."

And then maybe not.

you may also have similar beliefs about decision making. For example, a commonly held belief is that the world is "just" — people get what they deserve and deserve what they get. This is called the "just-world" theory;[36] people who believe this get into all kinds of trouble when they run up against situations that are not "just."

In the same way, belief in a one-right-answer gets you in trouble when you are comparing alternatives that vary in their desirability and amount of information available. You can be sure that, when you are trying to decide to stay in the military or not, a "one-right-answer" theory about the world will surely hang you up. Such a belief robs you of a whole range of good-enough alternatives. Believing there's only one right decision and all others are wrong is like playing Russian roulette with all the chambers of the gun loaded *except* one.

> *The sea endures no makeshifts. If a thing is not exactly right, it will be vastly wrong.*
> *— John Buchan*

5.2 B – THE DECISION POINT TRAP

The second biggest trap is to think of a decision as a "point" instead of a process.

Decision Trap Belief #2: Everything that precedes a point of decision is wasted energy.

Escape Hatch #2: Deciding is a process that takes place over time.

When you consider decision making as a point-in-time, you automatically frustrate yourself as you go through the normal back and forth of a decision-making process. This type of psychological trap leads to panic every time you feel like you've hit a dead end. Hitting dead ends is part of the process! They certainly provide information about directions you want to head away from. They can even provide some information if you reverse them.

Most of us lose some sleep over big decisions. This occurs because deciding is a process over time, not a neat little pinpoint in space. It's only with hindsight, when you have finished deciding something, that you can see what looks like a point-in-time, usually the time when you finished deciding. While you're in the middle of deciding, it's an ongoing process. Even when it seems that decisions are made quite quickly, it's usually because the homework of sorting through stages went on just under the surface of your awareness or the steps are so familiar that you run through them quickly and automatically, like deciding which way to drive to a familiar place.

Treating a decision as a process saves you from all the traps that go with trying to come up with a single right/wrong decision. Figure 5–2 walks you through such a process, and along the way you will encounter more beliefs that can turn into decision traps.

Much of the material about decision making as a four-step process is from two psychologists referenced earlier, Janis and Mann. They have looked closely at how people's actual decisions develop over time. They point out that most research on deciding focuses on cool rational models of decision making. Some of you got similar training from Uncle Sam on deciding on ways to get a job done. There are many such objective and mathematical tools for decisions in work settings. The problem is that in our own lives, we have many subjective feelings that don't fit these ways of deciding. When our feelings don't fit in with these strategies for cool thinking, we either try to ditch the feelings or just get frustrated and ditch the decision tool. What's needed is a way of including feelings in making personal decisions. Feelings (especially when they get hot) often point to critical information about our truest hopes for the future.*

Janis and Mann provide a realistic guide because they recognize that personal decision making is "hot" and that it occurs in real time. The clock is always ticking! The process includes all the fits;

---

* One Couple's Decisive Steps: Max always thought of himself as a practical man. He lived for flying and wasn't about to give up any of the chances the government gave him to fly at their expense. Sometimes, banking into a turn, he'd remember he was getting paid for this and break into a grin. If he looked over at his navigator, he'd usually get a grin back.

Jill didn't feel so great about Max's career. She knew if he didn't fly, he'd be a pretty miserable person, but she saw her husband as very unpractical. She figured he'd sell his soul to fly and felt that if he didn't make the shift to civilian flying in the next few years, his age would be against him. She wanted him to make the break while the economy was stable, not during an oil crunch when all those companies that needed pilots were cutting back. She'd bring it up to him every few months and get very little response. Meanwhile she didn't dare start career plans of her own since she kept expecting Max to start planning his change soon. She kept on with her volunteer work where she was well appreciated by the head of a social agency who said he hoped that her husband stayed stationed where he was forever.

— The Decision Process: For Max and Jill, the decision to re-enter civilian life took place over a three-week period; it took nine months to implement it. The steps were:

1. Max got angry and jealous because he didn't like the way Jill was dancing with her boss at a fund-raiser party on Saturday night.
2. They sniped at each other all week while Max flew extra flights and Jill put in extra volunteer hours.
3. The following Sunday, an old friend, Rick, called from the coast to tell Max to come get in on the expansion of a company out there.
4. A big fight followed, starting with Jill's "How come you never talk about getting out when I bring it up, but when Rick says, 'Come on,' now you're interested?" The fight ended with tears, making up, looking at a map of the states together and considering where they'd like to settle whenever Max got out.
5. The following Sunday, Max walked into the kitchen from the garage and said he wanted to start sending letters out just to see what jobs were available.

Nine months later, they were on the coast, settling into a new house. Max was part of a test team for a company that pays him well. It turned out to be a long way from the company where his friend Rick was, but a short commute for Jill to a program offering management training with a special emphasis on non-profit agencies.

starts, and stalls that go with any human feelings. It's frustrating to expect a decision about whether or not to leave the military to pop out like a total out of a cash register or toast out of the toaster. Hot models of decision making provide more tools to keep your frustration level down.

Janis and Mann[37] trace the decision process through a flow chart of questions (see Figure 5–2). You enter this flowchart by asking Question #1, "Are the risks serious if I don't change?" Here is where some of the hot thoughts and feelings come in. (In the preceding FYI box, Jill was worried that Max would lose civilian career choices with each birthday.) Risks include physical harm, being dishonorably discharged, ulcers, family stress, missing a great opportunity in the civilian sector, or any number of things. Notice that there are risks of negative events occurring and of positive events being missed. Usually some kind of negative feedback or some kind of civilian opportunity raises this first question.

For many, each birthday or anniversary of signing up for the military serves to remind you that you will not be able to stay in forever because the clock is ticking. When you hear about a possible opportunity, how does it move from the level of just shooting-the-bull about someday, to the level of a serious opportunity?

If you answered NO to Question #1, "Are the risks serious if I don't change?" — no risks of continued negative feedback, no serious loss of opportunity — then you are done with this section of the book. You can continue your decision process even though there's no current trigger for change. Read on and practice in preparation for the day those risks do become serious. If you haven't reached your third round of wondering about leaving the military, you can save your work so far and wait until it comes up a third time.*

If you answered YES to Question #1, go on to Question #2 in Figure 5–2, "Are the risks serious if I do change?" If you answer NO to Question #2, look where it puts you. If you're going to lose by not changing and there's no risk in changing, it seems you may be on your way to change for sure. You follow the arrow to the right, to the

> *Feelings sometimes make a better calendar.*
> — *Rod Steiger*

---

* The Rule of Three: When I work with parents, some of their stress is because they have no ground rules about when to take what their kids do seriously. One couple taught me the Rule of Three and I've passed it on to many people.

Like most useful rules of thumb, it's simple. If something bugs you once, ignore it. Twice…, forget it. It could be just a fluke, or a stage, or a lack of food or sleep or the flu. The third time — that's the time to say to yourself: THIS IS SERIOUS AND I'M GOING TO DEAL WITH IT!

If this is the third time you've considered when to leave the military, the Rule of Three suggests you conclude: THIS IS SERIOUS AND I'M GOING TO DEAL WITH IT! This also means that the risks of not changing are serious.

circle that says "No conflict — Change!" You may change your position in the military or you may be on your way out the Uncle Sam door, but you need to change something.

If you answered YES to Question #2, "Are the risks serious if I do change," you move down the vertical arrow and run directly into the hope question, Question #3. This asks, "Is it realistic to hope to find a better solution?" In other words, "Is there realistic hope that I can reduce the risks of either staying in or going by taking some action?"

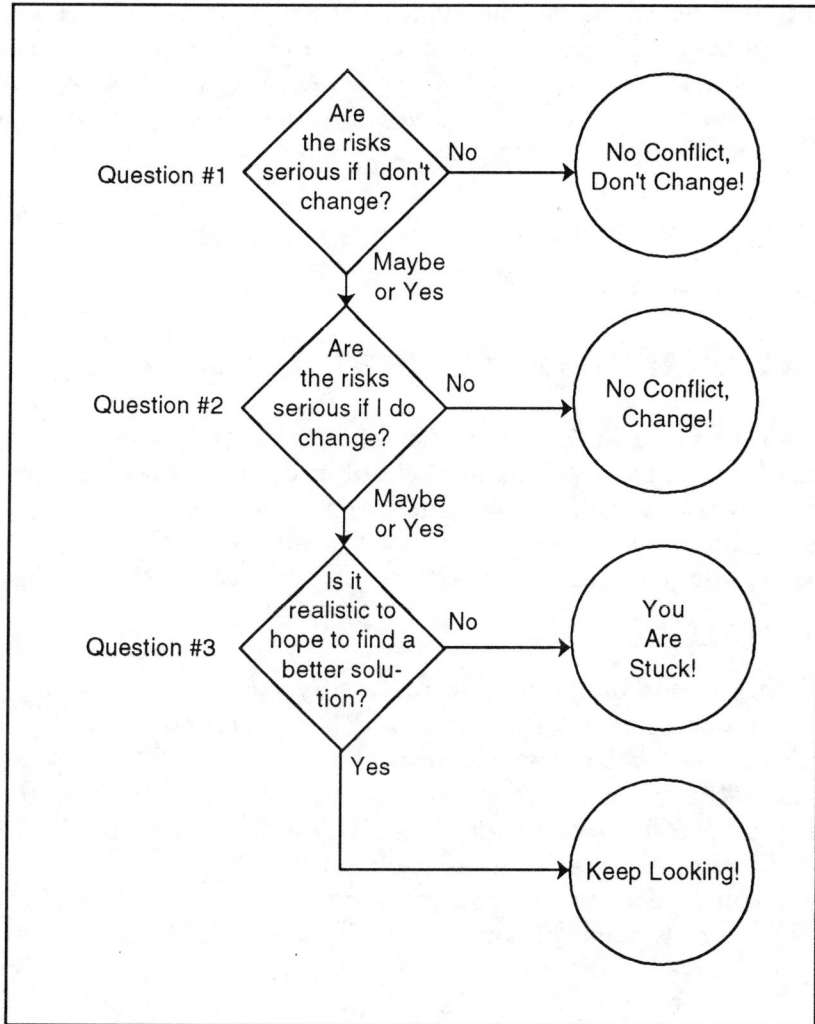

Figure 5–2: Flowchart of Change

If you feel that you can't reduce the serious risks of staying and can't reduce the serious risks of going, you are likely to stop right here. In the midst of hot personal decision making, when you don't see any hope of a better solution, you hang fire, you avoid the whole issue, you stop the decision making cold. You're stuck motionless (as described in Figure 5–1) and immobilized in a stall. Keep reading.

> *You must do the thing you think you cannot do.*
> *— Eleanor Roosevelt*

Looking back over your Lifemap decision points, you can see how many times you have experienced the kinds of traps described here. Have you avoided or regretted decisions afterwards because you were searching for the **right** decision rather than the **best** one? Have you made some decisions just to get out of being stuck in the decision-making process? Have you ended up stuck and wanting a change, but having no hope of a solution you could live with? Mark any of these Decision Traps on your Lifemap.

☆ ☆ REVIEW YOUR LIFEMAP ☆ ☆

5.3 – AVOIDANCE TRAPS

If you have answered NO to Question #3 in Figure 5–2, you are at the point where hopelessness about reducing risks can lead to avoiding decision making. When the problem is recognized, but there is no hope for a better solution, there are three styles of being stuck in avoidance: stalling, delegating, and bolstering.

5.3 A – STALLING

Stalling is one of the less creative ways to avoid the issue. You know all about stalling. It's that good old procrastination,* where you tell yourself all sorts of rationalizations: you'll "do it soon," there's something you "have to do first," it's "too late to start today," there's "plenty of time to think about it later." That kind of stalling. Most of the mental tricks and distractions to help you avoid the issue don't have any socially redeeming value at all. You're just kidding yourself. The only silver lining for confirmed avoiders is that, if you use physical tasks to distract yourself, then at least you might end up with a cleaned-out garage while you avoid decision making. Other

> *There are risks and costs to a program of action. But they are far less than the long-range risks and costs of comfortable inaction.*
> *— John F. Kennedy*

---

* The Procrastinator's Code: "Everything should go easily and without effort. It's safer to do nothing than to risk and fail. I can't afford to let go of anything or anyone. There is a right answer and I'll wait 'til I find it." [38]

advantages or silver linings are more psychological.* By doing nothing, you are protected from your fears of making a mistake, from fears of not making the *right* decision, and from fears of failure.

Scarlett O'Hara stalled in "Gone with the Wind." Scratching in the red earth after the Civil War had rolled over Tara, she chomps on a carrot and vows, "I'll think about that tomorrow." Well, fiddle-dee-dee! Sophisticated executives can also stall; they do it by using complicated graphs and charts to show how the "timing is not optimal" to make a risky decision. They can keep adding more lines and arrows to those risk charts rather than make a decision.

5.3 B – DELEGATING

More creative than stalling is delegation of the responsibility. Some people lay it off on caribou bones.

"When food is short because of poor hunting, the Labrador Indians consult an oracle to determine the direction the hunt should take. The shoulder blade of a caribou is put over the hot coals of a fire; cracks in the bones caused by the heat are then interpreted as a map. The directions indicated by this oracle are basically random, but if the Indians did not use a random number generator, they would fall prey to their previous biases and tend to over-hunt certain areas. Furthermore, any regular pattern of the hunt would give the animals a chance to develop avoidance techniques. By making their hunting patterns random, the Indians' chances of reaching game are enhanced.

"During extreme ambiguity, people feel out of control and there is a tendency toward incapacity — to do nothing. Under such circumstances, of course, it is more appropriate to do something — anything — since activity may uncover elements of control that were previously unnoticed. Thus, to the extent that superstitions give the feeling of control, they may encourage necessary activity.

"Also, in a random world, the best course of action is random action. Well-designed magical rites do precisely this — they encourage random action."[40]

Caribou bones aren't very easy to come by in our culture, so we're often reduced to apparently random techniques such as dice, the weather ("If it rains, I'll stay."), and palm-reading. Even if these

---

* **The Sweet Stalls - From the Diary of a Georgia Philosopher:** "It used to be that I couldn't stand (1) doing anything I was forced to do or (2) killing time waiting for someone else. I realized yesterday that I'm coming to treasure such moments as sweet islands of non-decision."[39]

don't work the way the caribou bones do, there can be a beneficial effect. Although you've left it up to the fates, the instant the coin leaves your hand you realize what you really want to do!

We also turn to other people to guide our avoidance and this process is not random. "If the detailer doesn't get me my preferred assignment, I'll really think about leaving." "When my wife's promotion comes through (my son makes the team, daughter is accepted to college, ex-wife gets married, etc.), we leave." Of course, one of the advantages to buck-passing is that it provides a scapegoat to blame for any of the problems that come out of a decision. Decision delegating is especially tempting when the military has been making most of your big decisions for years. When things have not gone well, there is some solace for military families in being able to blame the service, the assignment, or the guy up-the-line. You can turn to the military for information to make this decision about when to leave, but you won't be able to pin the blame for the final decision on them.

It helps to be able to compensate for any biases that might come with any information you get. Two basic questions help with this, (1) Is this person an expert or authority in this area? and (2) How likely is this person to be biased or shade the truth because they have a stake in or an opinion about my decision? If you rely on military input, you must judge the likelihood that this source can be objective about your possible civilian future as well as watch for mere opinion masquerading as valid information.

Go back over your change points on your Lifemap and take note of the times you:

- Leaned on someone in authority for a major life decision
- Utilized information from another while correcting for their biases.

☆ ☆ REVIEW YOUR LIFEMAP ☆ ☆

5.3 C – BOLSTERING

Bolstering is noticing only those plusses and minuses that support the decision you've made. Leon Festinger, a social psychologist,[41] has developed a theory which allows us to predict which kind of bolstering people will do under what kinds of conditions.

There are two types of bolstering. The first is where you magnify the positive aspects of your choice, sometimes called "enhancing." We've all experienced this. For example, once we choose a particular make of car, we begin to see that model everywhere; we notice, without effort, all the good features. You can hear yourself and your friends do the same kind of "enhancing" about their computers, VCRs, and other household purchases. People also use enhancing when they convert the negative aspects of their choice into a positive. For example, when a purchase is expensive, they might say, "It'll be good for me to learn to budget."

The second type of bolstering is "sour grapes," where you minimize the positive aspects of the Option you did not choose. One way people offset the positive aspects of the Option they are not choosing is to reduce its attractiveness: "I didn't really care about a view anyway; it would just be distracting."

Both types of bolstering help minimize conflicting feelings. They help us deal with the expectation that the Option chosen should be the "right" one with no negative aspects at all and the Options not chosen should be "wrong" ones, with all negatives and no positives.

You might notice that we've been talking about how bolstering occurs after the decision is made, but bolstering occurs both before and after decisions are announced. For example, patients choosing elective surgery were seen to "play up the necessity for the operation by imagining that their disorder might turn out to be cancer," even before their decisions were announced. Bolstering increases, however, after commitment has taken place in front of another person, as Festinger has shown.[42]

When someone makes a choice to maintain the status quo, in this case to stay in the military, there is a special kind of reverse bolstering that can occur. In addition to the normal bolstering of magnifying the good stuff and minimizing the hassles, they might also switch from minimizing problems to magnifying them. That is, we might embellish the problems associated with the choices we made. Reverse bolstering occurs when a family retells the stories of their worst camping disasters, just as they start out on this year's tent vacation. Festinger found this tendency even in white rats who would return to places where they got shocked most. He phrased this tendency as a sub-rule of bolstering: "You come to love those things for which you suffer." This natural tendency to reverse bolster with "war stories" helps you minimize the psychological impact of the known negative aspects of the status quo Option that you've elected.

Review your Lifemap decision points to identify those that might be connected to these kinds of psychological traps. Here are some questions to help you add notes to your Lifemap.

1. Note decisions that were plagued by more than your normal level of procrastination and stalling.

2. Note change points that you left up to other people to settle. Note also ones that you let ride on the roll of the dice (or the cracks in the caribou bones).

3. Note change points that you followed-up with bolstering. Record which were sour grapes for the Options you didn't take, super cheerleading for mediocre Options you chose, and "war stories" about the known problems when you chose to stick with the status quo.

☆ ☆ REVIEW YOUR LIFEMAP ☆ ☆

Being aware of these normal reactions to decision making may help you spot difficulties in coming up with your Number One Option. If you see yourself giving the Locus of Control for events to Chance or Powerful Others, you may need to spend more time to explore your reactions further. Note Options which evoke the urge to stall. See if you can identify a particular CON that you want to avoid, and then get outside help in finding ways to minimize its negative effects. Other patterns may require some non-rational maneuvers to break the logjam, such as:

•   You might wish that someone else would make your decision. Get a friend to help you design a poll so that you base your decision on a summary of friends' opinions.

•   You might wish that something else would make your decision. Spend time connecting with fate (dice, the I Ching, fortune tellers, psychic advisors) or higher powers (God, nature, other spiritual help) and write down what ideas you get and your reactions to them.

•   You might find that you are ready to go for a particular Option, except for some CONs that will be difficult for you. Do some bolstering ahead of time.

•   You might be having trouble giving up an Option. Prepare stories ahead of time to bolster how you gave up the PROs of the Option you're about to reject.

5.4 – TIMING TRAPS

Once you've skirted the dangers of hopelessness and the traps of stalling, delegating, and bolstering, you are faced with the question of whether there is sufficient time to search and deliberate among the alternative courses of action.

Is there time? If yes, you can then continue an orderly process of coming up with Options, revising and comparing them until you get to your best decision. But if the answer is, "No, there is no time." What can you expect then?

5.4 A – NOT ENOUGH TIME

When there is not enough time, there is a strong tendency to make a decision based on what Janis and Mann call hypervigilance. In their words, hypervigilance is: "...the pattern of overreacting to impending threat by taking impulsive, ill-considered action in a state of panic." In other words, jumping the gun. There are many ways to jump the gun, but they all leave you wide open for later regret. Some of these are:

1. Using an overly simple decision rule. For example, if the office gets new desks, stay — if not, go. Note that this rule is neither random (like caribou bones) nor related to factors that are vital to a job decision.

2. Doing what everybody else does. Calling around to your friends who have been in a similar situation, then letting the majority rule. In this case, you especially need to watch for situations where no one has any better information than you do. You could all check around to see if there's any danger in staying or going, and since no one sees anyone else moving, take that as evidence that not moving is the best thing to do. This has been called "pluralistic ignorance." It's used to explain why often no one will make the first move when there is danger in a crowd situation. Each person looks around to see what to do and, when they see no one else taking action, do nothing themselves.

3. Do what you did the last time you thought about leaving or re-enlisting. Keep doing what you've been doing, regardless. Basing life decisions on this rule of consistency or commitment means you are taking in no new information since the last time you decided. A good check on this tendency is to ask, "Would I make this decision again if I knew then what I know now?"

4. Finally, when there's not enough time to think things through, you can begin to focus your attention on very small details either of the job itself or job decision information. For example, taking exquisite care with a filing or retrieval system so that if you go you can use the same system on your next job, or, if you stay your files will be in great shape. Or, taking the ideas in this book and

making list after list of more and more details. It's a way of using detail work instead of a tranquilizer to stay calm.

Figure 5–3: Facing the Unknown

When people feel forced to make decisions in spite of not having anywhere near enough time, they tend to feel helpless. In job situations, most folk just sit tight, saying that they won't leave now under these rushed conditions, but promise themselves that they will carefully consider their alternatives after the time crunch is over. Most of us have experienced how easy it is to then forget the presence of any problems, settling back into our security blanket until the next crisis comes along.

5.4 B – TOO MUCH TIME OR INCHING YOURSELF INTO A CORNER

We've all been in situations where decisions have been reached by such small steps that you pass the point of no return without realizing it.* Sales people know this pattern all too well; Fuller brush folk and door-to-door sellers of aluminum pots, encyclopedias, and vacuum cleaners call it the foot-in-the-door technique.** Once you say "yes" to a small request ("May I have just a moment of your time?"), you are more likely to say "yes" to a second request, even if it is larger.

Another technique, the "hidden package deal," gets a group of people to agree to one part of the deal before unveiling the rest, and then no one wants to be the one to pull out first.[45] This whole area of making a commitment in front of other people is a source of decision traps. Stating something publicly, or to friends, often "freezes" a decision. Many folks have tried to use this kind of leverage on themselves to make difficult changes, like announcing to all your friends that you're quitting smoking. Although not the strongest tool for ending addictions, such public commitment certainly increases conflict if you relapse on your stated intention.

Another variation is called "commitment entrapment," where one slips from one action to another with subtle changes in self-image that make it easier and easier to do things you never would in the beginning. The psychological research done on willingness of people to administer electric shock to other people startled many. Researchers were unprepared to find that "nice folks" administered

---

* **Incremental Decision Making:** "One typical way in which people find themselves stuck with unwanted decisions is through a gradual, stepwise increase in commitment such that the final action, which would have been rejected if faced head-on, becomes a matter of 'Now-it's-too-late-to-get-out-of-it'."[43]

** **Foot in the Door:** Research data shows that if you first make a small request of people, they are more likely to agree to a second moderate request than if you just ask for the moderate request outright. In addition, if you ask people first for a large request, they are less likely to agree to a second moderate request as compared to making only one initial moderate request.[44]

stronger and stronger shocks in spite of the visual evidence of the pain they were causing. These tests caused a minor revolution by showing how people can be led to violate their basic values. (They also revolutionized the ethics of psychological experiments, so that studies can no longer be done that way.)

The same commitment entrapment process can occur when making repairs on an old car. You replace some small stuff because you're not quite ready to sell it. Each repair doesn't cost very much, but then, once those things are done, you can end up going from the $18.98 repair to the $137.50 one very quickly. Once you've got that much work in it, you begin to consider doing a major overhaul, even though, in the beginning, you promised yourself not to fix it up. Another example of this process is when you're put on hold on the phone. As you continue to hold, it gets harder and harder to hang up because you've already invested so much time in waiting.

One way to prevent such "creeping incrementalism" is to pre-set limits on how much energy, time and money you will spend on a particular goal, and tell someone else about these limits. Ask them to review with you how much energy, time and money has to go to other projects as well.

To apply this technique to staying on the phone when you're on hold, decide on a time limit ahead of time and tell someone about that limit. Be sure to look at the other demands on your time (or your stacked desk) when you're tempted to extend the limit, "...just a few minutes more. After all, I've already put in this much time, and I'll only have to call back anyway...etc., etc." Persons who operate small businesses at home often need this kind of structure so that they can choose if and when to go full time, rather than sliding into it without any real decision making.

A problem with incremental decision making is that we tend to do it when there is little time for thinking things through. We respond by making little decisions to "put out fires" and then these little decisions come to control our later decisions. When this style gets going, it's hard to go back and get a fresh look at the whole situation.

Accountants have a useful model for situations where prior commitments seem to leave us no alternative but sticking with what we've already invested in. They call such past investments "sunk costs," and the concept is useful for time and energy already spent as well as dollars. They suggest not to include these "sunk costs" in the next decision you make, since they are sunk in the past no matter whether you stay with what you've been doing or decide to strike out for something new. Accepting "sunk costs" as sunk and now irrelevant can keep people from being automatically stuck in staying with what they've been doing just because they've made an investment. Gamblers have always known this. They ignore sunk costs, calling it "cutting their losses."

5.4 C – TAKING ENOUGH TIME — VIGILANCE

If you have hope for a satisfactory solution and you have time to generate and compare alternatives, you can proceed with what Janis and Mann call the process of vigilance, a careful consideration of the risks associated with each alternative. Now that you know what some common decision traps are, you can tell the difference between being stuck and going through normal decision steps. If you spend a few days with memories about your early days in the military, do not yell at yourself for not having come to a decision yet or for getting lost in "memory lane." Those memories will all have information you can enter on your Lifemap. You may go off on what seems like a tangent for a while and dig out an old set of house plans, or dig into booklets you've been saving on running a self-sufficient farm or starting a body shop. I repeat: these side trips are all part of the process of vigilant decision making.

Now, depending on how you talk to family and friends, they may have a tough time tagging along down memory lane and ahead toward your new career dreams. Decision making is not especially neat, especially while still in process, and it has its own way of winding around until you come to the place where you can feel comfortable about the best decision for you.

Go back over the change points on your Lifemap and make note of the times you:

• Made a decision sooner than you wanted to

• Made a decision by inching yourself into a corner so you felt you couldn't back out.

☆ ☆ REVIEW YOUR LIFEMAP ☆ ☆

Looking at your Final Balance Sheet, see how the element of time plays into each of your Options. Are there some that would be more desirable if you could do them sooner? Later? If you have deadlines, where do they come from? What if you accepted that you just can't meet the deadline? How would that affect the PROs and CONs of each Option?

Redo your Final Balance Sheet one last time, being aware of the Decision Traps and some ways to escape from them. Re-read the last section of Chapter 4 (Final Decision) and choose your Option.

If you've been through the Four Steps of this book and are still stuck, with no hope of finding your best Option or reducing the risks of either staying or going, you are in that dreaded condition described in Chapter 3 — a CON/CON decision with few PROs in sight. You are also likely to be in Decision Stress and feeling depressed or anxious. If you've done all the exercises in this book, get the best outside help you can. Bring your notes and Lifemap and Balance Sheet with you.

If you've used your awareness of decision traps to choose your Option, read on for help in beginning to implement your choice.

CHAPTER 6

CARRYING OUT YOUR DECISION

Now you are ready to begin making lists of things to do to implement your Chosen Option. Depending on how long your timetable for carrying out this Option is, the implementation steps might extend over several weeks or several years. As you continue to think through the reality of these steps, you will probably encounter clearer information, recognize the need for additional steps, and run into barriers. During this process, it will be helpful to keep updating the entries for your chosen Option on the Final Balance Sheet.

The Final Balance Sheet can be used for contingency planning. As you move ahead with your Number One Option, update information and steps for one or more of your other Options, so they serve as up-to-date contingency plans. During times of stress, it helps to know that your contingency plans are current and right at your fingertips. Several good guides to contingency planning in making career decisions are listed in Appendix A.

Tracking what happens as you implement Option #1 will quickly alert you when you come up against an obstacle. It also minimizes the likelihood of being knocked off course by problems that arise. When a problem does arise, be ready to use the information you gain in overcoming it to improve your next step. Even when a problem really sets you back, using the information gained from this foul-up is important. Many self-help books suggest ways to turn mistakes and failures into motivation and information for success (Appendix A and information from Yeager* and Ringer in the following boxes).

Other difficulties may add to the normal obstacles in the way of implementing your chosen Option. Problematic thinking can create extra anxiety as the risks that accompany your choice become clearer to you. The most common types of problematic thinking are: post-decision regret, pass-fail self-grading, and double-standard bookkeeping.

---

* Chuck Yeager's A, B, and C Contingency Plans: "Without rpms there's no hydraulic pressure and without hydraulic pressure you can't move the stabilizer wings on the tail and without the stabilizer wings you can't control this bastard at the lower altitudes. He's in a steady-state flat spin and dropping... He's whirling around at a terrific rate... He makes himself keep his eyes pinned on the instruments... A little sightseeing at this point and it's vertigo and you're finished... He's down to 80,000 feet and the rpms are at a dead zero... He's falling 150 feet a second... 9,000 feet a minute... *And what do I do next?...* here in the jaws of the Gulp... *I've tried A! —I've tried B!* The damned beast isn't making a sound... just spinning around like a length of pipe in the sky... He has one last shot... the speed brakes, a parachute rig in the tail for slowing the ship down after a highspeed landing... The altimeter keeps winding down... Twenty-five thousand feet... but the altimeter is based on sea level... He's only 21,000 feet above the high desert... The slack's running out... He pops the speed brake... *Bango!* —the chute catches with a jolt... It pulls the tail up... He pitches down... The spin stops. The nose is pointed down. Now he only has to jettison the chute and let her dive and pick up the rpms. He jettisons the chute... and the beast heaves up again! The nose goes back up in the air!... It's the rear stabilizer wing... The leading edge is locked, frozen into the position of the climb to altitude. With no rpms and no hydraulic controls he can't move the tail... The nose is pitched way above 30 degrees... Here she goes again... She's back into the spin... He's spinning out on the rim again... He has no rpms, no power, no more speed chute, and only 180 knots airspeed... He's down to 12,000 feet... 8,000 feet above the farm... There's not a goddamned thing left in the manual or the bag of tricks or the righteousness of twenty years of military flying... Chosen or damned!... It blows at any seam! Yeager hasn't bailed out of an airplane since the day he was shot down over Germany when he was twenty... I've tried A!—I've tried B!—I've tried C!... 11,000 feet, 7,000 from the farm... He hunches himself into a ball, just as it says in the manual, and reaches under the seat for the cinch ring and pulls... He's exploded out of the cockpit with such force it's like a concussion... He can't see... *Wham...*a jolt in the back... It's the seat separating from him and the parachute rig... His head begins to clear... He's in midair, in his pressure suit, looking out through the visor of his helmet... Every second seems enormously elongated... infinite... such slow motion... He's suspended in midair... weightless...

"...Nearly down... He gets ready... Right out of the manual... A terrific wallop... He's down on the mesquite, looking across the desert, one-eyed... He stands up... Hell! He's in one piece!"

Yeager's fourth contingency plan — *D* — was the one that did the trick![46]

6.1 – POST-DECISION REGRET

Unforeseen problems and painful times will occur when implementing almost any Option except one that calls for no change. On the days when hassles mount up, you are susceptible to a big case of post-decision regret.

When you bolster your decision by emphasizing the positive, as discussed in Chapter 5, you're probably trying to avoid facing post-decision regret. There's no problem in regretting the costs and hassles that go with your Option but if this turns into a case of regretting *the decision itself*, you are likely to become depressed and anxious over what are normal costs of any decision. Hitting rough spots need not mean you made the wrong decision, nor that you need regret the decision itself.

The best way to avoid catching a bad case of this disease is by emotionally inoculating yourself with a series of "what-if" shots,* early in the implementation phase. Again, the references in Appendix A can help you transform your concerns about what might happen into contingency and back-up plans.

6.1 A – PASS-FAIL SELF-GRADING

The second way to do yourself in as you track your implementation steps is to fall into a pass/fail mentality. Inoculate yourself against this depressing habit by deciding now what you would take as a sign of failure. The number of problems encountered? Too slow in carrying out implementation steps? Coming up against unpredicted problems?

---

* What-ifs as Emotional Inoculation against Post-Decision Regret:
1. "What if we borrow the money to make the move, and one of our parents becomes dependent on us because of illness?"
2. "What if I'm half-way ready to change military programs and they phase out the one I'm headed for?"
3. "What if I get the civilian job I want, and then when we've resettled I get fired or laid off?" Whatever your brain can come up with for worries and concerns, write them down faithfully, especially in the beginning of the implementation phase, and later turn them into contingency and backup plans, like:
1. Check on parents' health coverage now. Add plans for them, including getting lower monthly payments on a loan so you can also build an emergency fund or pay insurance premiums.
2. Grill, grill, and grill your career counselor so you know the financial and career Options that are left if your military track gets derailed.
3. Talk with family about possibility of being fired or getting laid-off and come up with a tighten-the-belt plan that's kept on the back burner until probation is over.

The new information you get, as you live through the experiences your choices bring, quickly turns into hindsight. It can become a special kind of guilt: How Come I Didn't Know? Getting angry because you did not know this information ahead of time may provide a lesson as well as its frustration. Is there something you should have done earlier to find out this information? Can you now take those information-gathering steps? This is a chance to use Ringer's Sustenance of a Positive Attitude ideas.* If you have a lifelong style of blaming yourself for "not knowing better," you may not be able to change it in the middle of this career decision. You can, however, use the information that you now *do* know, once it is clear to you and once you stop kicking yourself. Use it to recalibrate your course as you move onward.

6.1 B – DOUBLE STANDARDS FOR LOCUS OF CONTROL

Remember the Locus of Control styles: Internal, External/ Chance, and External/Powerful Others? As you experience the outcomes from the implementation steps of your Option, notice to which style you attribute Locus of Control. People sometimes attribute the source of control differently for good news and bad news, rather than thinking the Locus of Control is the same whether the outcome is good or bad.

---

* Using Foul-ups to Sustain a Positive Attitude: Robert Ringer has a Theory of Sustenance of a Positive Attitude through the Assumption of a Negative Result. This theory talks about "...admitting in advance that most things don't work out due to factors beyond my control, knowing that I'll lose battles but not the war and be prepared not to win each time out.

"While you are enduring all of life's bumps and bruises in a daily stream of lost battles, there is a freebie in each situation. It comes in the form of an educational experience. All you need to do is extract the lesson learned and use it to be better prepared the next time around. The negative result itself is history. Forget it. The knowledge gained, however, can prove to be far more valuable than a victorious battle. It sounds paradoxical, but in reality the 'sustenance' theory prepares you for long-term success by preparing you mentally for short-term failure.

"...This mental preparation is another form of price paying. You must face the inescapable reality that absolutely everything in life has a price: love, friendship, material gain, a relaxed mind, the freedom to come and go as you please — anything which adds pleasure to your existence. All things worth obtaining must be paid for. If you delude yourself into thinking otherwise, you only open the door to endless frustration."

Odd as it sounds, it's also helpful to pre-think what you are going to do if you are met with overwhelming success, especially if you start a small business in the civilian world that has the possibility of catching on fast. Ringer would call this Sustenance of a Positive Attitude through the Assumption of a Positive Result that you May Not Be Ready For. Manufacturers of new items have to live with this malicious twist of Murphy's Law when they are unable to accept large orders from national retailers because they did no planning for big-time success.[47]

Some double standards build you up. If things turn out well, you might say, "I did great" (Internal). If things turn out poorly, you could think, "Lousy timing" (External/Chance), or "They ruined it for me" (External/ Powerful Others).

Other double standards tear you down. If things turn out well, you might say, "They made it possible" (External/Powerful Others), or "What a lucky break" (External/Chance). If things turn out poorly, you might take it all on yourself, "I really blew it" (Internal). Notice if you tend to follow one of these patterns, but realize that you can change any one of these patterns with practice.

6.1 C – DECISION FLIPPING

Some people, under certain conditions, can have a strong reaction to a small piece of information and then make sudden, strong decision reversals without warning. They literally "flip" to a different viewpoint, accompanied by a sudden reversal of their previous stance.* If this has already happened to you, it probably led to periods of frustration during decision making. Whenever you suspect this is happening, grab your Final Balance Sheet to refocus on The Big Picture.

6.2 – THE FINAL TOUCH

There are no other tricks to mucking through the implementation phase of decision making, but there is plenty of practical

---

* **The Catastrophe Theory of Outcomes:** Rene Thom,[48] mathematician, proposes a catastrophic model[49] of how flipping back and forth between Options in sudden jumps and reversals can occur. One application to dogs' behavior states that given at least a moderate tendency toward fear (A) and aggressiveness (B) in the dog, (think of your moderate tendencies to Option A and Option B on your Balance Sheet), it takes only a slight change in the incoming stimulus pattern to result in a sudden and dramatic outcome (hence the idea of catastrophe). When the dog is showing both fear (ears back) and rage (teeth bared), rarely does a neutral behavior occur. Rather, a *slight* increase in fear sends the dog running off or a *slight* increase in rage, even in an already fleeing animal, results in a sudden attack. The math gets complicated but provides some nice predictive details. First is the idea that once the animal is at a certain level of fear and rage, neutral behavior is very unlikely. Also the specific order of buildup of fear and rage (which came first, last, and how many switches) influences outcomes and that, although a *slight* change can result in a catastrophic outcome, once the leap is made, it is only reversible by a very much larger change in the stimulus conditions.

In terms of your tendencies toward two of your Options, this theory would say that once both tendencies are at least moderate, it may take only a *slight* increase in one to topple you over toward one or the other. Which one depends not only on the input but on the sequence of prior leanings toward one and the other. Once you have toppled in one direction, it will take much more input to reverse that direction than it did to get you over there in the first place. Note that the theory predicts you will not be in a neutral state.

guidance which focuses on how to be smart about carrying out your decision. See Appendix A for various kinds of help from books and other resources.

You have now made a decision about whether to stay in or to muster out to civilian life, or you know you need more help to do it. This decision may be modified by events or you may even move to a back-up Option or contingency plan. You may continue to use the forms and information you've compiled here or not need lists to get on to your next step. Either way, take a minute to update your Lifemap by marking in your Numero Uno Option in the right spot. Perhaps five or ten years from now you will enjoy looking back and seeing how your life has gone on from this point on your Lifemap.

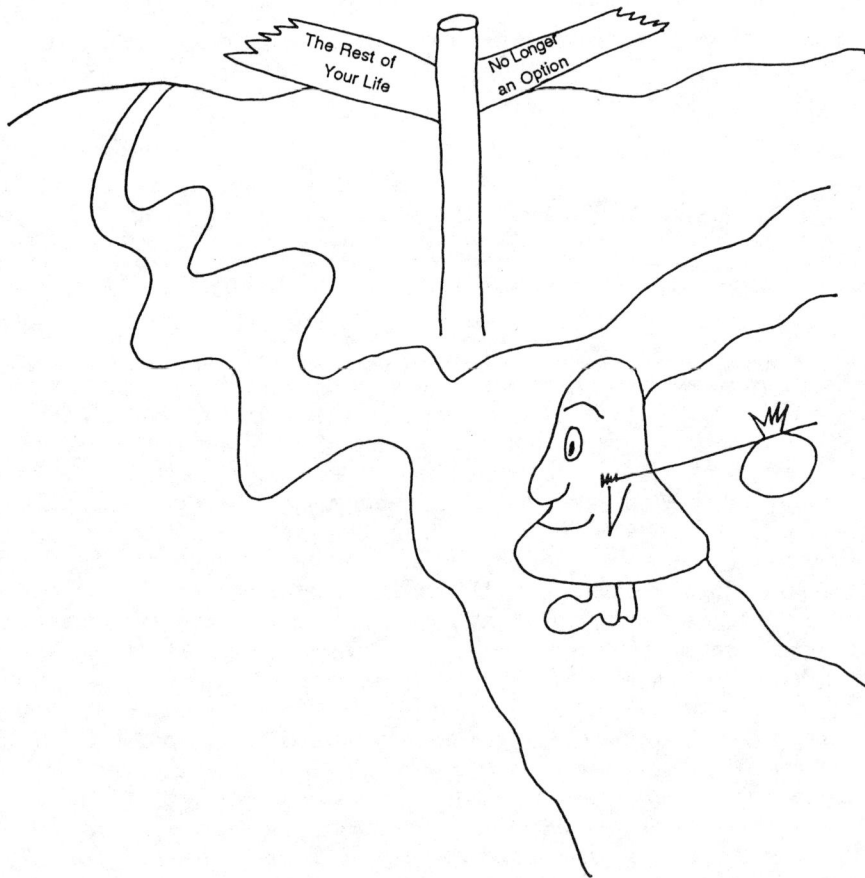

Figure 6–1: The Rest of Your Life

NOTES

1   Progoff, Ira. *At a Journal Workshop.* New York: Dialogue House, 1975, p. 134.

2   Progoff (1975), p. 133.

3   Mayer, Nancy. *The Male Mid-Life Crisis: Fresh Starts after 40.* New York: Doubleday, 1978, p. 193.

4   Rahe, R. H. Life change and subjects' subsequent illness reports. In E. K. E. Gunderson and R. H. Rahe (eds.), *Life Stress and Illness.* Springfield, Ill.: Charles C. Thomas, 1974.

5   Rahe (1974), pp. 134-163.

6   Zimmerman, M., O'Hara, M. W., and Corenthal, C. P. Symptom contamination of Life Event Scales, *Health Psychology,* 1984, *3* (1), 77-81.

7   Mayer (1978), p. 63.

8   Kobasa, S. C. Hilker, R. R., and Maddi, S. R. Who stays healthy under stress? *Journal of Occupational Medicine,* 1979, *21* (9), 595-598.

9   Savage, R. E., Perlmuter, L. C. and Monty, Richard A. Effect of reduction in the amount of choice and the perception of control on learning. In Perlmuter, L., and Monty, R. *Choice and Perceived Control.* Hillsdale, N.J.: Lawrence Erlbaum, 1979.

10 Glass, D. C. and Singer, J. E. *Urban Stress: Experiments on Noise and Social Stressors.* New York: Academic Press, 1972.

11 Seligman, Martin E. P. and Miller, Suzanne M. The psychology of power: concluding comments. In L. Perlmuter and R. Monty (Eds.) *Choice and Perceived Control*. Hillside, N.J.: Lawrence Erlbaum, 1979.

12 Seligman, Martin E. P. *Helplessness: On Depression, Development, and Death*. San Francisco: Freeman, 1975.

13 Harvey, John H. Attribution of freedom. In Ickes, A., Harvey, J. H., and Kidd, R. F. *New Directions in Attribution Research (Vol. 1)*. Hillsdale, N.J.: Lawrence Erlbaum, 1976.

14 Hall, Calvin S. and Lindzey, Gardner. Stimulus-Response Theory. In Hall, C. S. and Lindzey, G. (Eds.) *Theories of Personality*. New York: John Wiley, 1970.

15 Willing, Jules Z. *The Reality of Retirement*. New York: William Morrow, 1981, p. 195.

16 Willing (1981), Chapter 5.

17 Bazerman, M. H. Why negotiations go wrong. *Psychology Today,* 1986 (June).

18 Skinner, B. F. *Beyond Freedom and Dignity*. New York: Bantam/Vintage, 1971, p. 40.

19 Anon. *From the Diary of a Georgia Philosopher*. Unpub. ms., 1984.

20 Janis, Irving L. and Mann, Leon. *Decision Making: A Psychological Analysis of Conflict, Choice, and Commitment*. New York: The Free Press, 1977, p. 136.

21 Kahneman, Daniel and Tversky, Amos. Choices, values, and frames. *American Psychologist*, 1984, *39*, 341-350.

22 Janis and Mann (1977).

23 Deal, Terence E. and Kennedy, Allan. *Corporate Cultures; the Rites and Rituals of Corporate Life*. Reading, Mass.: Addison-Wesley, 1982; especially Chapter 6.

24 Brehm, J. W. and Cohen, A. *Explorations in Cognitive Dissonance*. New York: John Wiley, 1962.

25 Baker, Nancy C. *Act II. The Mid-Career Job Change and How to Make It*. New York: Vanguard Press, 1980, p. 205.

26 Bolles, Richard N. *The Three Boxes of Life and How to Get Out of Them: An Introduction to Life/Work Planning*. Berkeley, Cal.: Ten Speed Press, 1978, p. 363-364.

27 Bolles (1978), p. 362.

28 Baker (1980), p. 43.

29 Bolles (1978), especially Chapter 2.

30 Levenson, H. Distinctions within the concept of internal-external control: Development of a new scale. *Proceedings of the 80th Annual Convention*, American Psychological Association, 1972.

Rotter, J. B. Generalized expectancies for internal versus external control of reinforcement. *Psychological Monographs,* 1966, *80,* 1.

Perlmuter L. and Monty, R. *Choice and Perceived Control.* Hillsdale, N.J.: Lawrence Erlbaum, 1979.

31 Friedrich, James R. Perceived control and decision making in a job hunting context. *Basic and Applied Social Psychology.* 1987, *8,* 163-176.

32 Friedrich, James R. A perceived control analysis of decision-making behavior. (Doctoral Dissertation, University of Michigan, 1984) *Dissertation Abstracts International,* 45, 2994B.

33 Hoff, Benjamin. *The Tao of Pooh.* New York: Penguin, 1982, pp. 85-87.

34 Marino, Kenneth E. and White, Sam E. Departmental structure, locus of control and job stress: the effect of a moderator. *Journal of Applied Psychology,* 1985, *70,* 782-784.

35 Progoff (1975), p. 135.

36 Lerner, M. J., Miller, D. T., and Holmes, J. Advances in Experimental Psychology. New York: Academic Press, 1976.

37 Janis and Mann (1977), pp. 196-197.

38 Burka, Jan B. and Yeun, Lenora M. *Procrastination.* Reading, MA: Addison-Wesley, 1983.

39 Anon. (1984).

40 Gimple, Martin L. and Dakin, Stephen R. Voodoo strategies. *California Management Review,* 1984, *27,* 1.

41 Janis and Mann (1977), p. 82.

42 Janis and Mann (1977), p. 83.

43 Janis and Mann (1977), p. 287.

44 Cialdini, Robert B. *Influence : How and Why People Agree to Things.* New York: William Morrow, 1984.

45 Janis and Mann (1977), p. 295.

46 Wolfe, Tom. *The Right Stuff.* New York: Bantam, 1979, pp. 358-361.

47 Ringer, Robert. *Looking Out for Number One.* Los Angeles, CA: Fawcett Crest/CBS, 1977, pp. 25-26.

48 Thom, Rene. *Structural Stability and Morphogenesis.* Reading, Mass.: W. A. Benjamin, 1975.

49 Zeeman, E. C. Catastrophe theory. *Scientific American,* 1976, (April), 65-83.

APPENDIX A - ANNOTATED BIBLIOGRAPHY

The large number of books on careers decisions are of two types: general books that claim to cover a lot of material (the only-book-you-will-ever-need) or more specific books that zero in on one specific area or skill (e.g.,resumés for communication jobs). When the titles are not obvious, the mini-review usually makes it clear how much breadth of information your getting.

1. CAREER PLANNING - GENERAL

Bolles, Richard N. *The 1992 What Color is Your Parachute: a Practical Manual for Job-Hunters and Career Changers*. Berkeley, Cal.: Ten Speed Press, 1992. <u>The</u> classic guide to knowing yourself through questionnaires that lead to specific goal setting and career plans. See also *How to Create a Picture of Your Ideal Job or Next Career* which has the main questionnaires from "Parachute" in an 8 1/2" x 11" format, or *The Quick Job-Hunting (And Career-Changing) Map* in a 5" x 8" format.

Bolles, Richard N. *The Three Boxes of Life*. Berkeley, Cal.: Ten Speed Press, 1981. Directions to free yourself from the typical life stages of learning (school), working, and, then, playing (retirement). Lots of exercises to arrange your life in the combinations you want. Good for opening up new ideas and directions.

Bolles, Richard N. *How to Find Your Mission in Life.* Berkeley, Cal.: Ten Speed Press, 1991. Spiritual approach to finding your deep life focus.

Fowler, Elizabeth M. *New York Times Career Planner.* New York: Times Books, l987.

2. CAREER PLANNING - FOR THE MILITARY OR THE JOB YOU ALREADY HAVE

Jaffe, Dennis T. *Take this Job and Love It: How to Change Your Work without Changing Your Job.* New York: Simon and Schuster, 1988. Recipes for handling burnout at your company. Good for checking your stress levels and clarifying your own "mission." All civilian examples.

Komar, John J. *The Great Escape from Your Dead-End Job.* New York: Ballantine, 1980. Guide to revitalize the job you have or change; uses civilian corporate examples.

Lloyd, Joan. *The Career Decision Planner: When to Move, When to Stay, and When to Go Out on Your Own.* New York: John Wiley, 1992. Excellent review of job dissatisfactions and ways to understand company cultures. Useful if you can translate to military cultures.

McKay, Robert. *Planning Your Military Career.* Lincolnwood, Ill.: VGM Career Horizons, 1984. Overview of all five services for those deciding to enlist or not . A dated review to remember how you got in and how the services have changed since then.

Potter, Beverly A. *Beating Job Burnout.* New York: Ace Books, 1980. All examples based on corporations. Good for those in military management or administration.

3. GENERAL CAREER CHANGE

Banning, Kent B. and Friday, Ardelle. *How to Change Your Career.* Lincolnwood, Ill: VGM Career Horizons, 1991. (Skip 1987 version, Planning Your Career Change). One worksheet on "Career Planning for Involuntary Re-entry." Good guide, although not specifically for military, covering resumés, interviews, salary negotiation.

Cross, Andrea S. G. *Planning a New Strategy for Mid-Life.* New York: Crown, 1991. Overview. Focus on money, family, dreams, goals. Charts organize information.

Hyatt, Carole. *Shifting Gears: How to Master Career Change and Find the Work That's Right for You.* New York: Simon & Schuster, 1990. Good overview of job change due to changes in the world. Four styles of handling work (Lifer, Builder, Synthesizer, Reinventor) useful for seeing a variety of jobs as part of one pattern.

Johnson, William Courtney. *The Career Match Method.* New York: John Wiley, 1992. Specific steps, clear and even fun, using new concepts of matching.

Krannach, Ronald L. *Careering and ReCareering for the 1990's: The Complete Guide to Planning Your Future.* Manassas, Va.: Impact, 1989.

Shingleton, John D. and Anderson, James. *Mid-Career Changes: Strategies for New Direction in the 90's.* Orange, Cal.: Career Publishing, 1992.

Strasser, Stephen and Sena, John. *Transitions: Successful Strategies from Mid-Career to Retirement.* Hawthorne, N.J.: Career Press, 1992.

4. LEAVING JOBS

Barranger, Jack. *Knowing when to Quit.* San Luis Obispo, Cal.; Impact Publishers, 1988. Excellent analysis of the leaving aspect of your IN or OUT decision. Helps guard against going or staying for the wrong reasons.

Gale, Barry & Gale, Linda. *Stay or Leave: A Complete System for Deciding Whether to Remain at Your Job or Pack Your Traveling Bag.* New York: Perennial /Harper and Row, 1989.

Holloway, Diane and Bishop, Nancy. *Before You Say "I Quit": a Guide to Making Successful Job Transitions.* New York: Collier Books/MacMillan Publishing, 1990. Forms to fill out; only one chapter on decision making.

Lefkowitz, Bernard. *Breaktime: Living without Work in a Nine-to-Five World.* New York: Penguin, 1980. Tips for handling isolation and other problems of not working.

Levenson, Jay Conrad. *Earning Money without a Job.* New York: Henry Holt, 1991.

5. FROM MILITARY TO CIVILIAN

Fitzpatrick, William G. and Good, C. Edward. *Does Your Resumé Wear Combat Boots? Successful Transition from Military to*

Civilian Life: A Job-Seeker's Guide. Charlottesville, Va.: Blue Jeans Press, 1990. Clear overview of job-hunting steps of networking, resumés, interviews, negotiating. Includes examples of translating military experience into general skills and a timetable.

Henderson, David G. *Job Search: Marketing Your Military Experience in the 1990's.* Harrisburg, Pa.: Stackpole Books, 1991. More checklists and overall organizers than Fitzpatrick and Good.

Lee, W. Dean. *Beyond the Uniform. A Career Transition Guide for Veterans and Federal Employees.* New York: John Wiley, 1991. Similar to Henderson. and Fitzpatrick and Good but more chatty and more emphasis on getting through the transition itself, including financial planning.

Nyman, Keith O. *Re-Entry: Turning Military Experience into Civilian Success.* Harrisburg, Pa.: Stackpole Books, 1990. This second edition still provides a good overview of the issues, based on a one-year pre-separation timetable. More a discussion than workbook format.

6. JOB HUNTING

The order of events here is networking for contacts, resumés, interview skills, follow-up and negotiation. Guides that focus only on the civilian climate may help you tune to civilian styles. Several authors have written series (Allen, Yates, etc.).

Beatty, R. H. *The New Complete Job Search.* New York: John Wiley, 1992. Good overview.

Farr, J. Michael. *The Very Quick Job Search.* Indianapolis, IN: JIST Works, Inc. 1991. Good. Includes how to make your own JIST cards as a tool for effecive networking; quick specific steps in an overall approach.

Jackson, Tom. *Guerrilla Tactics in the New Job Market.* New York: Bantam, 1991. Clear, crisp, one-stop job hunting guide.

Petras, Kathryn and Petras, Ross. *The Only Job Hunting Guide You'll Ever Need.* New York: Poseidon Press, 1989. Well laid out.

Networking

Armstrong, Howard. *High Impact Telephone Networking for Job Hunters.* Holbrook, Mass.: Bob Adams, 1992. Up-to-date techniques not included in most other books.

<u>Resumés</u>

Allen, Jeffrey G. *The Perfect Reference.* New York: John Wiley, 1990. Gives specifics and sample form letters.

Beatty, Richard. H. *175 High Impact Cover Letters.* New York: John Wiley, 1992. Samples to borrow from for the letter that goes with the resumé.

Beatty, Richard H. *The Resumé Kit.* New York: John Wiley, 1991. Directions and lots of examples.

Bostwick, Burdette. *Resumé Writing (4th Edition).* New York: John Wiley, 1990. Big, includes good guidelines for handling a mailing list. Samples of ten styles of resumés.

<u>Interviewing</u>

Allen, Jeffrey G. *Get the Interview.* New York: John Wiley, 1990. An almanac of many, many techniques. Not so good at helping you plan which ones to use where but a good source of ideas.

Allen, Jeffrey G. *The Complete Q and A Job Interview Book.* New York: John Wiley, 1988. Format allows reader to go right to problem areas. Includes a good personal inventory section and how to prepare using a cassette recorder.

Drake, John D. *The Perfect Interview: How to Get the Job You Really Want.* New York: AMACOM, 1991. Very detailed preparation of specific interview skills and answers. Unlikely that just reading it will be of help, but actually doing his exercises can't help but build confidence and improve your interview presentation.

Yates, Martin. *Knock 'em Dead with Great Answere to Tough Interview Questions.* Holbrook, Mass.: Bob Adams, 1992. Good practice material, e.g. "How would you evaluate me as an interviewer?" Also in series — Resumés That Knock 'em Dead, Cover Letters That Knock 'em Dead.

<u>Negotiation</u>

Allen, Jeffrey G. *The Perfect Follow-up Method to Get the Job.* New York: John Wiley, 1992.

Chapman, Jack. *How to Make $1000 a Minute: Negotiating your Salaries and Raises.* Berkeley, Cal.: Ten Speed Press, 1987.

7. SPECIAL SPOTS

Benjamin, Janice and Block, Barbara with Jones, Kathryn. *How to be Happily Employed in Dallas-Fort Worth.* New York: Random House, 1990. A good general guide of job change steps, the same for each area covered. Also in series is Boston, San Francisco, Washington, DC.

Camden, Thomas M. and Steinberg, Sara. *How to Get a Job in Seattle/Portland*. Chicago, Ill.: Surrey Books, 1990. Includes general how-to chapters and then lists of information for each city. Also in series: Atlanta, Chicago, Dallas, Ft. Worth, Europe, Houston, Los Angeles/San Diego, New York, San Francisco, Washington, D.C.

Cetraon, Marvin. *The Great Job Shake-Out. How to Find a New Career after the Crash*. New York: Simon and Schuster, 1988. One professional forecaster's predictions of where and where not to head with your career.

Farr, J. Michael. *America's Top Technical and Trade Jobs*. Indianapolis, Ind.: Jist Works, 1992. Discusses overall trends as well as specifics for each job; excellent use of tables to organize and compare complex information.

Farr, Michael J. and Martin, Kathleen. *America's 50 Fastest Growing Jobs*. Indianapolis, Ind.: JIST Works, 1991. More down-to-earth specifics from the JIST folks.

Hoover, Gary, Campbell, Alta, and Spain, Patrick (Eds.) *Hoover's Handbook of American Business*. 1992. Austin, Tex.: The Reference Press, Inc. 1991. Researching the companies you want interviews with.

Kleinman, Carol. *The 100 Best Jobs for the 1990's: General Trends and Beyond*. Chicago, Ill.: Dearborn Financial Publishing, 1992.

Lauber, Daniel. *Government Job Finder*. River Forest, Ill.: Planning Communications, 1992. Good if you know you want this sector; listings by industry and by state. Military areas are broadcasting, dieticians, logistics engineers, or military club personnel.

Morgan, Hal and Tucker, Kerry. *Companies that Care: the Most Family-Friendly Companies in America — What They Offer and How They Got That Way*. New York: Fireside/Simon Schuster, 1991. Case studies of companies from personal perspectives.

Petra, Kathryn and Petras, Ross. *Jobs '92*. New York: Poseidon Press, 1992. Laid out by industry, less developed listings by career and region.

Smith, Carter. *America's Fastest Growing Employers: The Complete Guide to Finding Jobs with over 700 of America's Hottest Companies*. Holbrook, Mass.: Bob Adams, 1992.

8. SPECIAL PEOPLE

Anthony, R. J. and Roe, G. *Handling Ageism: Over 40 and Looking for Work?* Holbrook, Mass.: Bob Adams, 1991. General job hunting advice includes discussions of being unhappily employed because of age issues, review of discrimination guidelines, specific resumé and interview techniques for older job hunters.

Bastress, Frances. *The Relocating Spouse's Guide to Employment.* Chevy Chase, Md.: Woodley Publications, 1989.

Basta, Nicholas. *Environmental Jobs for Scientists and Engineers.* New York: John Wiley, 1992. Very specific, organized by occupation in an easy-to-use format.

Rivera, Miquela. *The Minority Career Book.* Holbrook, Mass.: Bob Adams, 1991.

Zeitz, Baila and Dusky, Lorraine. *The Best Companies for Women.* New York: Pocket Books, 1988. A catalog of 50 companies.

9. RETIREMENT

Doyle, Bob et. al. *Can You Afford to Retire?* Chicago, Ill: Probus Publishing, 1992. Excellent, specific, all the way the decision through to estate planning.

Savageau, David. *Retirement Places Rated.* New York: Prentice Hall, 1990. A catalog of well-organized listings.

Sharff, Lee E., Borden, E. and Stein, Fred. *Veteran's Benefits Handbook.* New York: Prentice Hall, 1992. Probably worthwhile to have your own copy of the basics.

10. WORKING FOR YOURSELF

Barnett, Frank and Barnett, Sharon. *Working Together: Entrepreneurial Couples.* Berkeley, Cal.: Ten Speed Press, 1988.

Hall, Daryl Allen. *1001 Businesses You Can Start from Home.* New York: John Wiley, 1992. Mental shopping through a catalog of enterprenurial possibilities.

Knight, Brian. *Buy the Right Business at the Right Price.* Manchester, Vt.: Country Business,1992. Lots of forms to fill in to make financial comparisons.

Kamoroff, Bernard. *Small Time Operator.* Laytonville, Cal.: Bell Springs Publishing, 1992.

11. STRESS

Anderson, Robert A. *Stress Power: How to Turn Tension into Energy.* New York: Human Sciences Press, 1978.

Benson, Herbert. *The Relaxation Response.* New York: Avon, 1975.

Davis, Martha, Eshelman, Elizabeth, and McKay, Matthew. *The Relaxation and Stress Reduction Workbook. Third Edition.* Oakland, Ca.: New Harbinger Publications, 1988.

12. DECISIONS

Cammaert, Lorna P. and Larsen, Carolyn C. *A Woman's Choice: A Guide to Decision Making.* Champaign, Illinois: Research Press, 1979.

Williams, Andrea. *Making Decisions.* New York: Zebra Books, 1985. Good guide for any decision. Teaches use of stick figures and clustering tools that anyone can draw plus an intuitive type of list making for those who hate organized detail.

APPENDIX B - LOCUS OF CONTROL SCALE
LEVENSON'S INTERNAL (I), POWERFUL OTHERS (P) AND CHANCE (C) SCALES

Below is a series of attitude statements. Each represents a commonly held opinion and there are no right or wrong answers. You will probably disagree with some items and agree with others. We are interested in the extent to which you agree or disagree with such matters of opinion.

Read each statement carefully. Then indicate the extent to which you agree or disagree by circling the appropriate number in front of each statement. The numbers and their meaning are indicated in the box at right:

First impressions are usually best in such matters. Read each statement, decide if you agree or disagree and the strength of your opinion, and then circle the appropriate number in front of the statement. *Give your opinion on every statement.*

If you find that the numbers to be used in answering do not adequatedly indicate your own opinion, use the one which is closest to the way you feel.

If you *agree strongly:*	circle +3
If you *agree somewhat:*	circle +2
If you *agree slightly:*	circle +1
If you *disagree slightly:*	circle −1
If you *disagree somewhat:*	circle −2
If you *disagree strongly:*	circle −3

	AGREE			DISAGREE			
1)	+3	+2	+1	−1	−2	−3	Whether or not I get to be a leader depends mostly on my ability.
2)	+3	+2	+1	−1	−2	−3	To a great extent my life is controlled by accidental happenings.
3)	+3	+2	+1	−1	−2	−3	I feel like what happens in my life is mostly determined by powerful people.
4)	+3	+2	+1	−1	−2	−3	Whether or not I get into a car accident depends mostly on how good a driver I am.
5)	+3	+2	+1	−1	−2	−3	When I make plans, I am almost certain to make them work.
6)	+3	+2	+1	−1	−2	−3	Often there is no chance of protecting my personal interest from bad luck happenings.
7)	+3	+2	+1	−1	−2	−3	When I get what I want, it's usually because I'm lucky.
8)	+3	+2	+1	−1	−2	−3	Although I might have good ability, I will not be given leadership responsibility without appealing to those in positions of power.
9)	+3	+2	+1	−1	−2	−3	How many friends I have depends on how nice a person I am.
10)	+3	+2	+1	−1	−2	−3	I have often found that what is going to happen will happen.
11)	+3	+2	+1	−1	−2	−3	My life is chiefly controlled by powerful others.
12)	+3	+2	+1	−1	−2	−3	Whether or not I get into a car accident is mostly a matter of luck.
13)	+3	+2	+1	−1	−2	−3	People like myself have very little chance of protecting our personal interests when they conflict with those of strong pressure groups.
14)	+3	+2	+1	−1	−2	−3	It's not always wise for me to plan too far ahead because many things turn out to be a matter of good or bad fortune.
15)	+3	+2	+1	−1	−2	−3	Getting what I want requires pleasing those people above me.
16)	+3	+2	+1	−1	−2	−3	Whether or not I get to be a leader depends on whether I'm lucky enough to be in the right place at the right time.
17)	+3	+2	+1	−1	−2	−3	If important people were to decide they didn't like me, I probably wouldn't make many friends.
18)	+3	+2	+1	−1	−2	−3	I can pretty much determine what will happen in my life.
19)	+3	+2	+1	−1	−2	−3	I am usually able to protect my personal interests.
20)	+3	+2	+1	−1	−2	−3	Whether or not I get into a car accident depends mostly on the other driver.
21)	+3	+2	+1	−1	−2	−3	When I get what I want, it's usually because I worked hard for it.
22)	+3	+2	+1	−1	−2	−3	In order to have my plans work, I make sure that they fit in with the desires of people who have power over me.
23)	+3	+2	+1	−1	−2	−3	My life is determined by my own actions.
24)	+3	+2	+1	−1	−2	−3	It's chiefly a matter of fate whether I have a few friends or many friends.

Scoring: Add your scores for the three sets of items below and then add 24 to each total.

1. _____	3. _____	2. _____
4. _____	8. _____	6. _____
5. _____	11. _____	7. _____
9. _____	13. _____	10. _____
18. _____	15. _____	12. _____
19. _____	17. _____	14. _____
21. _____	20. _____	16. _____
23. _____	22. _____	24. _____
_____ + 24 = _____	_____ + 24 = _____	_____ + 24 = _____
Internal	External/ Powerful Other	External/ Chance

Scores range from 0 to 48.
Average scores for a group of 90 adults were: Internal = 35.93, External/Powerful Others = 16.25, External/Chance = 13.96

APPENDIX C - FORMS

1. LIFEMAPS

2. TABLE OF OPTIONS

3. TRIAL BALANCE SHEET

4. TOP THREE ITEMS

5. FINAL BALANCE SHEET

1930's

New York World's Fair
War of the Worlds broadcast
Austria annexed by Hitler
Joe Louis wins title
FDR elected 2nd term
Jesse Owens wins Gold Medals
Huey Long assassinated

1935

Dionne quintuplets born
Benny Goodman's first band
1st All-Star baseball game
Prohibition repealed
Lindbergh kidnapped
FDR elected first time
Empire State Building completed
Depression deepens, banks close

Medical
Family/Friends
Residence/Travel
School/Jobs
Age

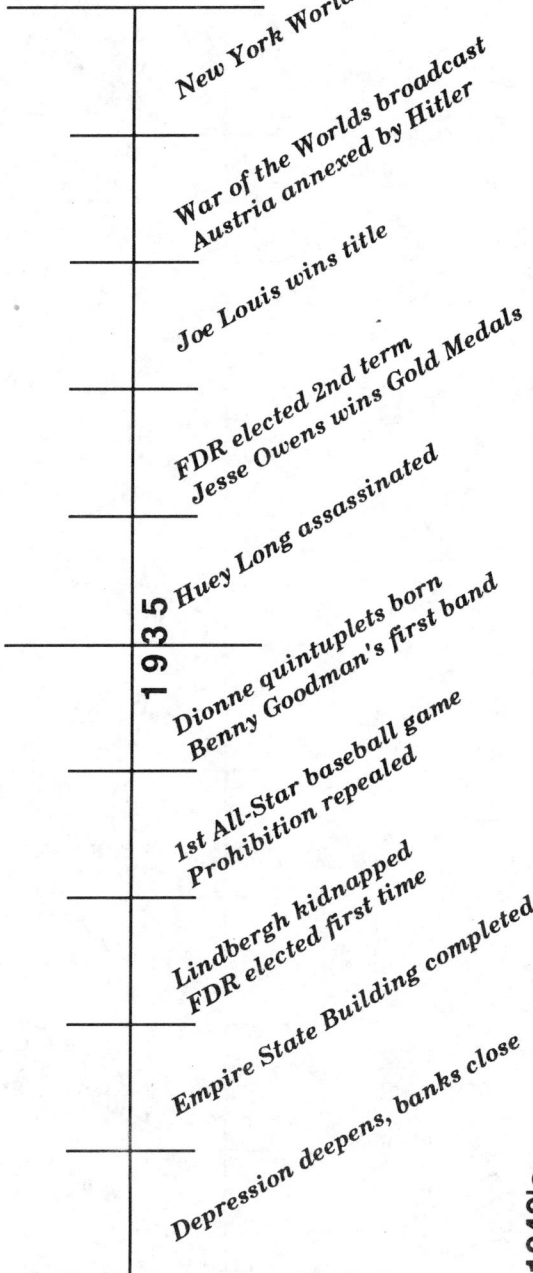

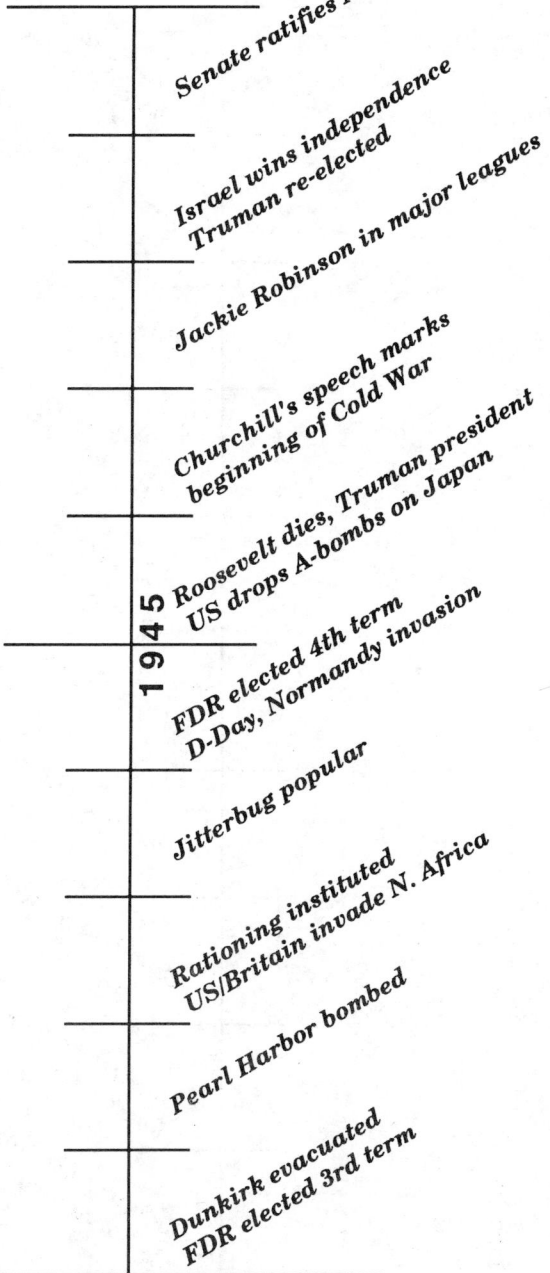

1940's

Senate ratifies NATO
Israel wins independence
Truman re-elected
Jackie Robinson in major leagues
Churchill's speech marks
beginning of Cold War
Roosevelt dies, Truman president
US drops A-bombs on Japan

1945

FDR elected 4th term
D-Day, Normandy invasion
Jitterbug popular
Rationing instituted
US/Britain invade N. Africa
Pearl Harbor bombed
Dunkirk evacuated
FDR elected 3rd term

Medical
Family/Friends
Residence/Travel
School/Jobs
Age

1950's

Age

- N. Korea invades S. Korea
- Kefauver hearings
- MacArthur removed from Korean Command
- Eisenhower elected
- Fighting ends in Korea
- McCarthy hearings
- James Dean killed
- Eisenhower re-elected
- Elvis Presley's Heartbreak Hotel
- Civil Rights Act passed
- NASA established
- Alaska and Hawaii become states

1 9 5 5

Medical
Family/Friends
Residence/Travel
School/Jobs

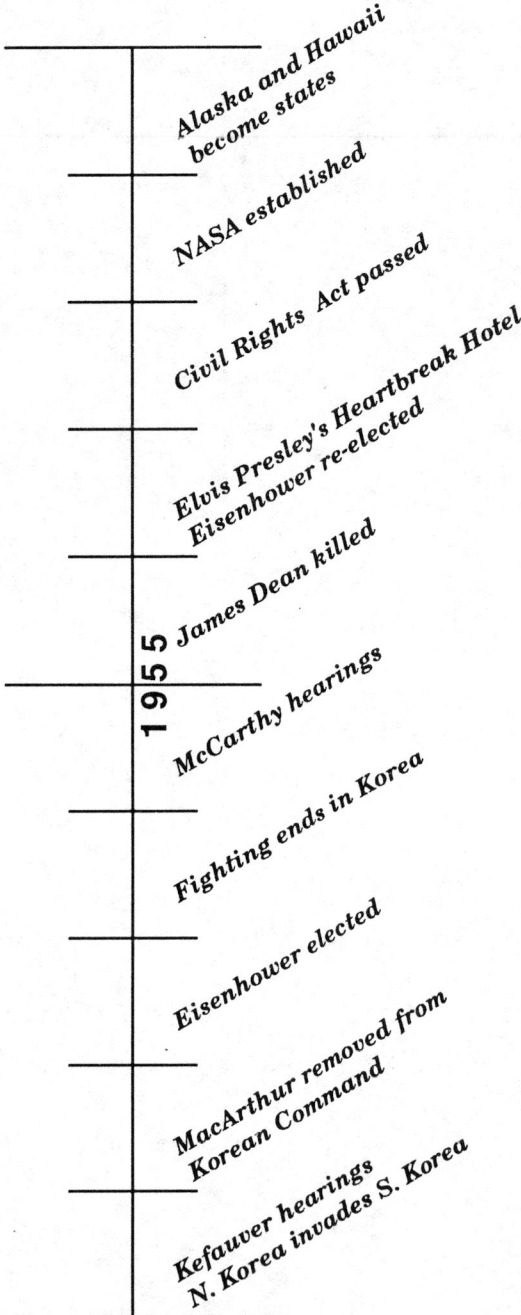

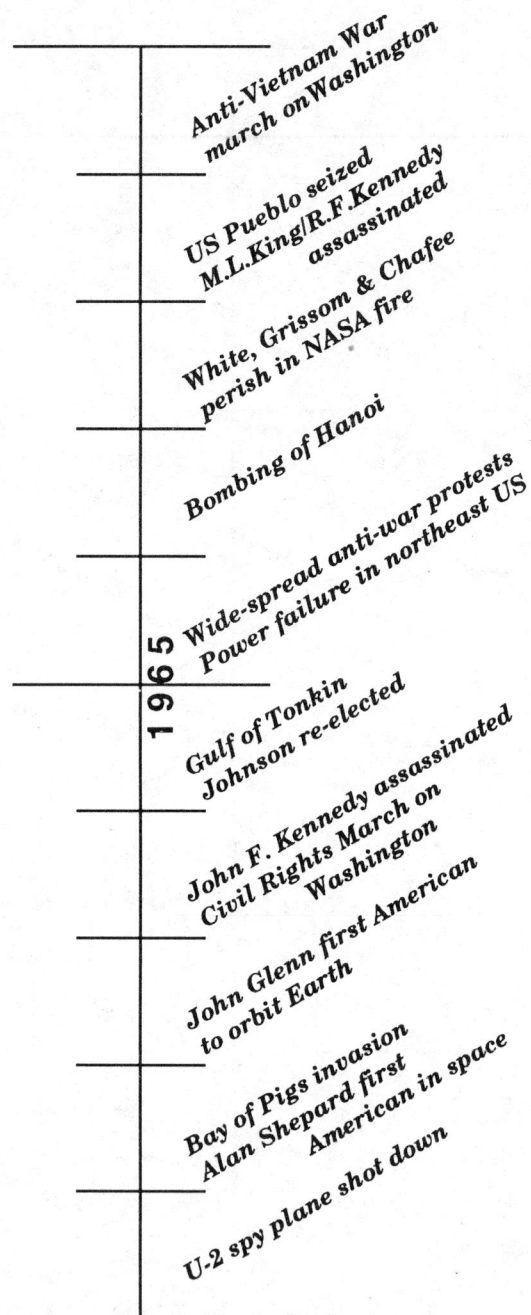

1960's

Age

- U-2 spy plane shot down
- Alan Shepard first American in space
- Bay of Pigs invasion
- John Glenn first American to orbit Earth
- Civil Rights March on Washington
- John F. Kennedy assassinated
- Johnson re-elected
- Gulf of Tonkin
- Power failure in northeast US
- Wide-spread anti-war protests
- Bombing of Hanoi
- White, Grissom & Chafee perish in NASA fire
- M.L.King/R.F.Kennedy assassinated
- US Pueblo seized
- Anti-Vietnam War march on Washington

1 9 6 5

Medical
Family/Friends
Residence/Travel
School/Jobs

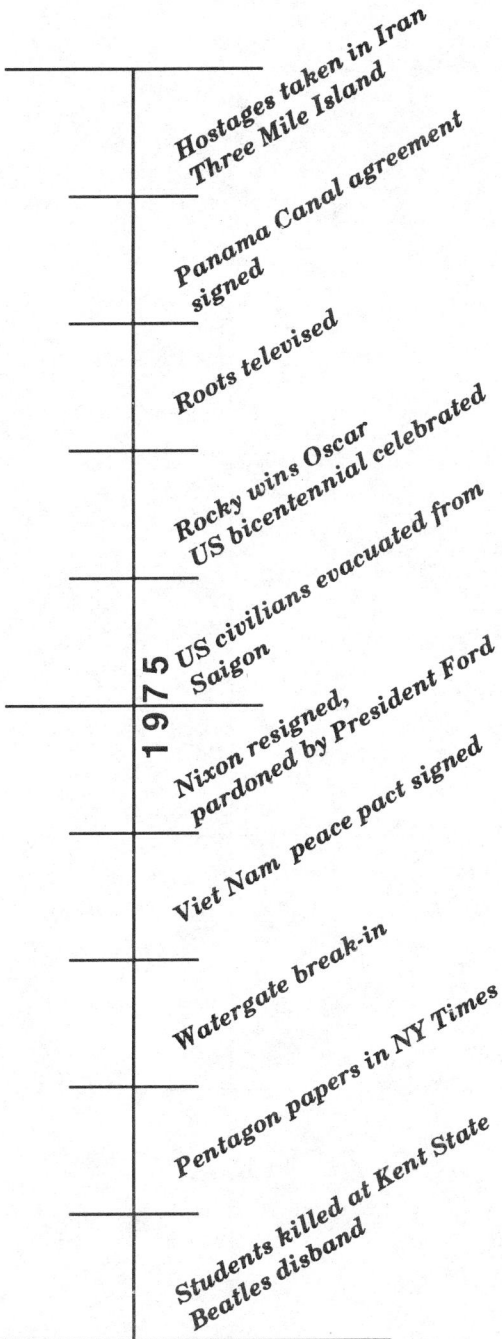

1970's

- Medical
- Family/Friends
- Residence/Travel
- School/Jobs
- Age

1975

Hostages taken in Iran
Three Mile Island

Panama Canal agreement signed

Roots televised

Rocky wins Oscar
US bicentennial celebrated

US civilians evacuated from Saigon

Nixon resigned, pardoned by President Ford

Viet Nam peace pact signed

Watergate break-in

Pentagon papers in NY Times

Students killed at Kent State
Beatles disband

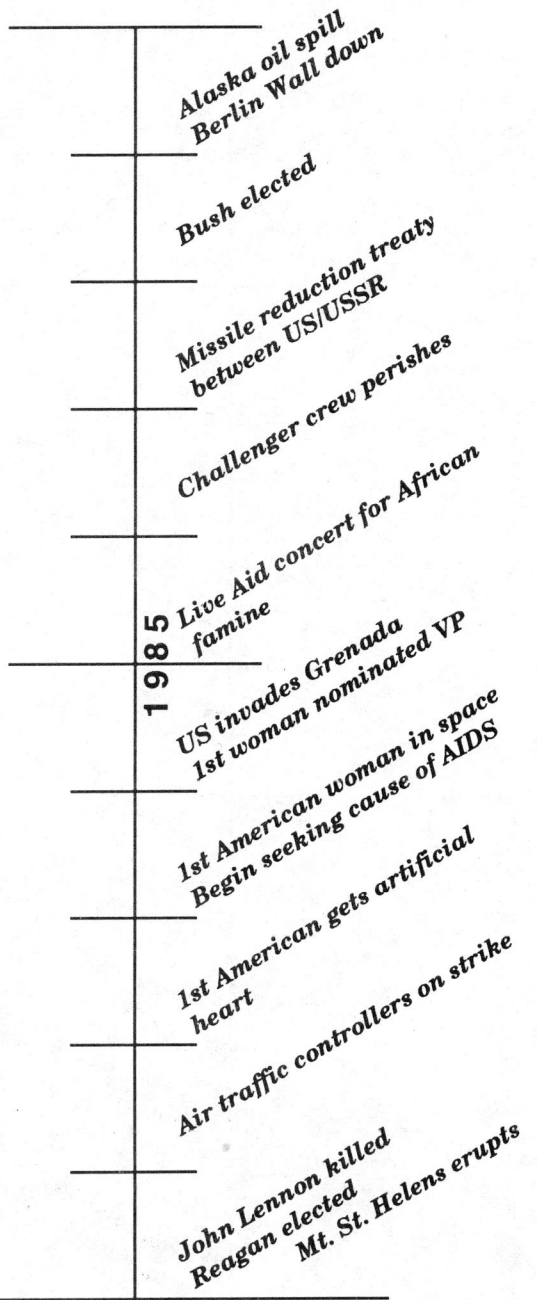

1980's

- Medical
- Family/Friends
- Residence/Travel
- School/Jobs
- Age

1985

Alaska oil spill
Berlin Wall down

Bush elected

Missile reduction treaty between US/USSR

Challenger crew perishes

Live Aid concert for African famine

US invades Grenada
1st woman nominated VP

1st American woman in space
Begin seeking cause of AIDS

1st American gets artificial heart

Air traffic controllers on strike

John Lennon killed
Reagan elected
Mt. St. Helens erupts

1990's

Medical

Family/Friends

Residence/Travel

School/Jobs

Age

1995

Breakup of USSR
Rodney King verdict

Gulf War   End of Cold War
Mt. Pinatubo erupts

Iraq invades Kuwait
Smoking banned on all US flights

2000's

Medical

Family/Friends

Residence/Travel

School/Jobs

Age

2005

TABLE OF OPTIONS

OPTIONS	AMOUNT OF CHANGE	DEGREE OF PREDICTABILITY	AMOUNT OF LEAVETAKING
1.			
2.			
3.			
4.			

TRIAL BALANCE SHEET

ANTICIPATED OUTCOMES	OPTIONS							
	1.		2.		3.		4.	
	☆ PRO ☆	☆ CON ☆	☆ PRO ☆	☆ CON ☆	☆ PRO ☆	☆ CON ☆	☆ PRO ☆	☆ CON ☆
1. Practical Outcomes for You								
A. Income								
B. Job security								
C. Difficulty of work								
D. Stress								
E. Interest level of work								
F. Chance to use old skills								
G. Chance for advancement or learning new skills								
H. Room for making own decisions								
I. Chance to fulfill goals outside of work								
J.								
2. Self Approval/Disapproval								
A. Hating quitters								
B. Meaning of the work								
C. Links to long-term goals								
D. Need to compromise								
E. Creativity of work								
F.								
3. Practical Outcomes for Others/Family								
A. Income								
B. Status								
C. Time for family and others								
D. Decision making stress								
E.								
4. Others' Approval/Disapproval								
A. Prior commitments								
B. Status								
C. Independence vs. recognition								
D.								

IDENTIFY THE TOP THREE ITEMS

Now fill in the following three sentences using the top three of your eight most important items (this is an extra form for the exercise in 4.1D):

The *first* most important thing to me is to

       get _____ (+3 points)

   or to avoid _____ (−3 points)

             an issue which is practical _____

        or one of approval/disapproval _____

                          for me _____

                        or others _____

The *second* most important thing to me is to

       get _____ (+2 points)

   or to avoid _____ (−2 points)

             an issue which is practical _____

        or one of approval/disapproval _____

                          for me _____

                        or others _____

The *third* most important thing to me is to

       get _____ (+1 point)

   or to avoid _____ (−1 point)

             an issue which is practical _____

        or one of approval/disapproval _____

                          for me _____

                        or others _____

FINAL BALANCE SHEET

	OPTIONS							
	1.		2.		3.		4.	
ANTICIPATED OUTCOMES	☆ PRO ☆	☆ CON ☆	☆ PRO ☆	☆ CON ☆	☆ PRO ☆	☆ CON ☆	☆ PRO ☆	☆ CON ☆
1. Practical Outcomes for You								
A. Income								
B. Job security								
C. Difficulty of work								
D. Stress								
E. Interest level of work								
F. Chance to use old skills								
G. Chance for advancement or learning new skills								
H. Room for making own decisions								
I. Chance to fulfill goals outside of work								
J.								
2. Self Approval/Disapproval								
A. Hating quitters								
B. Meaning of the work								
C. Links to long-term goals								
D. Need to compromise								
E. Creativity of work								
F.								
3. Practical Outcomes for Others/Family								
A. Income								
B. Status								
C. Time for family and others								
D. Decision making stress								
E.								
4. Others' Approval/Disapproval								
A. Prior commitments								
B. Status								
C. Independence vs. recognition								
D.								

INDEX

IN OR OUT
OF THE MILITARY

HOW TO MAKE
YOUR OWN BEST DECISION

D. F. REARDON, PH. D.

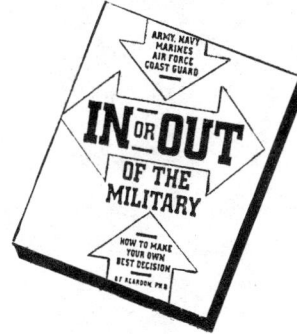

Chapter 1. Step 1 — Taking Stock
Chapter 2. Step 2 — Laying Out Your Options
Chapter 3. Step 3 — Trial Balance
Chapter 4. Step 4 — Summing Up
Chapter 5. Decision Traps and Escape Hatches
Chapter 6. Carrying Out Your Decision

In or Out of the Military is an objective decision-making guide to help you compare your military and civilian options and come to a conclusion about staying in or getting out of the military. Easy-to-use tools help you —

☆ See how life experiences shape your options
☆ Lay out the full range of your options and probable outcomes
☆ Deal with challenge and stress - stop stalling
☆ Weigh and compare your options for the decision that's right for you

Whether you decide to stay in or leave the military, you will be more confident of your decision after you have considered all the key issues and anticipated all the probable outcomes.

Author, psychologist, and teacher Dr. Diane Reardon has specialized in counseling active duty and retired military personnel and their families since 1977. Combining this experience with research breakthroughs, she has developed an effective process to help you make the most critical decision of your military life.

4 EASY WAYS TO ORDER!

1. BY PHONE TOLL-FREE 1-800-678-5519 Please have your VISA or MasterCard ready. M-F, 8-6; S 10-4, Pacific
2. FAX (206) 679-9191 Fax completed order form, 24 hours a day. Credit card orders only please.
3. THROUGH YOUR LOCAL BOOK STORE In or Out of the Military: How to Make Your Own Best Decision by Dr. Diane Reardon. 140 page trade paperback; Illustrations, Tables, Notes, Bibliography, Index. ISBN 1-882287-44-4, LC No. 92-061683. $14.95 plus shipping. Pepper Press, 1254 West Pioneer Way, Suite A266, Mail Order Dept. B, Oak Harbor, WA 98277.

4. CLIP AND MAIL THIS ORDER FORM TODAY!
GUARANTEE: If for any reason you are not satisfied with your purchase, simply return the undamaged book within ten days for a complete refund.

Ask about our quantity discounts

YES! Please enter my order for **IN OR OUT OF THE MILITARY: HOW TO MAKE YOUR OWN BEST DECISION**

_____ Copies @ $14. 95_____

SHIPPING AND HANDLING By surface mail book rate (2 weeks) $3.50 per book_____
I can't wait. Here is $2 more per book for priority mail (5 days to any US or APO/FPO address)_____

SUBTOTAL_____
Washington Residents please add 7.8%_____
TOTAL DUE_____

NAME_____
ADDRESS_____
CITY_____ STATE_____ ZIP_____
OFFICE_____ AUTHORIZED BY_____ P. O._____

METHOD OF PAYMENT
_____ VISA [VISA] _____ MasterCard [MasterCard] _____ Check or Money Order payable to Pepper Press
Card #_____
Expiration Date_____ Signature_____

Mail to • PEPPER PRESS   *Thank you!*
1254 West Pioneer Way
Suite A266, Mail Order Dept. B
Oak Harbor, WA 98277-3288

Others who might like to order **IN OR OUT OF THE MILITARY: HOW TO MAKE YOUR OWN BEST DECISION** are:
Name_____ Name_____
Address_____ Address_____
City _____ State_____ Zip____ City _____ State_____ Zip____
_____ Friend _____ Librarian___Counselor___Chaplain _____ Friend _____ Librarian___Counselor___Chaplain